V. K. Frederiks   A. A. Friedmann

Foundations of the Theory of Relativity

Volume 1: Tensor Calculus

With a Foreword by Alexander P. Yefremov

Translated by Svetla Kirilova-Petkova
and Vesselin Petkov

Edited by Vesselin Petkov

MINKOWSKI
Institute Press

V. K. Frederiks (1885-1944)
A. A. Friedmann (1888-1925)

ISBN: 978-1-927763-26-1 (softcover)
ISBN: 978-1-927763-27-8 (ebook)

Minkowski Institute Press
Montreal, Quebec, Canada
http://minkowskiinstitute.org/mip/

For information on all Minkowski Institute Press publications visit our
website at http://minkowskiinstitute.org/mip/books/

V. K. Frederiks
1885-1944

A. A. Friedmann
1888-1925

Foreword

Russians use to say: a man given a talent becomes talented in all. This observation fully fits to Alexander Friedman, a modest person but great personality, a man who lived a short life but left an undeletable trail. Born in the family by soul and profession devoted to music, the young Alexander had unexpectedly chosen not less beautiful but may be more challenging art of mathematics; we can just guess whether music, this math and harmony in sounds, could have influenced his unconscious choice.

However he perfectly succeeded from the very beginning: his first math study was awarded by a gold medal of the St. Petersburg University. Later he plunged into the mathematical jungle of hydrodynamics trying to describe the processes in the Earth's atmosphere, and to make himself sure that the processes really take place, he personally travelled in the skies in dirigibles and balloons.

Acquainted with the air, he as a volunteer went to the First World War, joined the just born Russian air forces, and several times participated in operational flights and aerial reconnaissance. He became a pilot himself and even was an aviator trainer.

After the war Friedmann, not strong in health, left the big cities, Kiev, St. Petersburg, Moscow, for the provincial Perm University (West Urals region) to work as a prosperous professor of mathematics (mechanics) and not less successful administrator, the university's vice-rector.

But the composition of tangled differential equations in partial derivatives and the search for their solutions was in all periods of Freidmann's life his main and imperishable passion. It has happened so or maybe it was a reflection of the sky-flight experience, but meteorology and atmospheric processes became his narrow field of scientific interest. Sophisticated models of compressible fluid fluxes and curls demanded to keep up an acquaintance with freshly developing math of differential manifolds and tensor calculus, but it would be in vain to look for solutions without hard work and brilliant intuition.

Happily Friedmann possessed all that. Easily absorbing new trends in different spheres of rapidly developing technology and science, he as a bright mathematician even faster coped with math difficulties and, not less importantly, penetrated into the notional mysteries of the Einstein's general relativity (GR); though it can be clearly exiv

plained: geometrized gravity was the closest analogue of the continued medium modeling physics of the Earth's atmosphere. Of course, the four-dimensionality of space-time and the indefinite character of its metric were exotic features of the new theory, but Friedmann "unpacked the box".

Utterly interested, Friedmann did his best to have an eye on publications devoted to GR and he surely was aware of Einstein's solution describing a static universe balancing between total attraction and − repulsion inevitably, so artificially, inserted into the theory equations in the form of cosmological Λ-term. One may argue, but it is plausible that the introduction of the Λ-term to some extent reflects the psychological conservatism of Einstein who, in contrast to his revolutionary steps in physics, was not able to throw down the chains of traditional thinking. In particular, his long-life skepticism about quantum mechanics is well known. Probably this conservatism forced him to look for "eternal", hence static, universe "once and forever" established by Isaac Newton a quarter of millennium earlier. But the static world suitable to the requirements of GR unexpectedly appeared to be limited in volume, not infinite, as that of Newton. Later De Sitter suggested another GR solution describing "static universe", but this cosmological model had even more surprising property: this world appeared to be empty, and the space-time curvature was determined by the "cosmological" Λ-term. Moreover, soon it was shown that both static models of the Universe, those of Einstein and De Sitter, were unstable under small perturbations.

We can just guess why Friedmann decided to search for dynamic cosmological solutions. Either he was emotionally moved by some global breaths of the Earth's atmosphere, as he conceived them studying meteorology, or maybe it was his specific personal response to great dynamics of the world at the beginning of XX century – with cruel social revolutions and powerful impetus overwhelmingly and impatiently pushing the human civilization to technical and scientific progress.

This great dynamics of the local Earth-centered domain of the Universe sharply sensed by Friedmann was in strong contrast with the deadly static character of Einstein's cosmological model. Moreover, one could suspect that from Freidmann's viewpoint the artificial Λ-term saving the static "universe", and not approved otherwise, should be thought of as an irritating factor just making the sophisticated system of the GR equations even more complex and hard to solve.

There was though heavy scientific reason to doubt the static character of the Universe. Not later than in 1914 the American astronomer Vesto Slipher detected great velocity, about 1000 km/s, of the Andromeda nebula motion relatively to the Sun, hence to our Galaxy. Moreover, soon he came to conclusion that all observed galaxies move off the Sun. Arthur Eddington, a man who in fact made famous GR and his author, immediately suggested the idea of an expanding world. It is hard to say if Friedmann had exhaustive information of all these facts, but he knew well that all cosmic bodies were moving. The world should be dynamic!

So once Friedmann decided to take chance. I just imagine him sitting at the table with a blank sheet of paper and thinking of a better ansatz for the metric tensor, thoroughly writing down the Christoffel symbols and curvature components, then adjusting them to get the Ricci tensor, scalar curvature and lastly Einstein's conservative tensor. Probably less effort took the choice of the energy-momentum tensor, since not long before Einstein was brave enough to take the galaxies for point-like particles uniformly filling the space. Friedmann followed the suit, and his Universe as well was conceived as a dust cluster homogeneously spreading in the isotropic space without compression though interacting gravitationally, i.e. subject only to the force of attraction. But in difference with Einstein's model Friedmann's "dust" could move, so a sufficient value of initial velocity of the galaxies guaranteed the existence of a dynamic Universe, at least for a certain time. In fact, it was the first time when the GR equations comprising two partial derivatives were considered for a solution.

And the solution was found! Strangely enough, it described not a unique model but simultaneously three types of non-stationary worlds: two models linearly and hyperbolically expanding with time, and one model harmonically pulsating. Moreover, it turned out that no Λ-term was needed for this beautiful solution, though possessing an initial singularity. This last factor in fact was left for future analysis. Published in *Zeitschrift für Physik* in 1922 this solution is known to have evoked skepticism of Einstein, but later the "genius of relativity" publicly recognized a mistake in his calculations.

The only question that remained was: if the Universe was indeed dynamic? The answer came in 1929. Having analyzed the correspondence between distances and red-shifts of remote galaxies Edwin Hubble discovered that they were interconnected by a linear function. The farther is distance of a galaxy, the greater is the speed of its flight. The world was really expanding.

But Friedmann already could not received this good news and celebrate the victory. In 1925 for the last time he raised to the stratosphere in a balloon, reached the record height of 7.5 km, and after successful landing suddenly felt bad. In a month he died.

Near a century went by since. More accurate observations revealed deviations of the linear law of galaxies separation thus demanding to include inflationary phases in the cosmological history. In turn it entails a revision of the total mass of Universe and the need of forces of repulsion with speculations of their sources, dark matter and dark energy. In a way all this caused Λ-term resurrection in the GR equation. Many variants were suggested to prevent the initial cosmological singularity. It is normal, the theories change with new data and new minds. But Alexander Friedmann will stay a man who "made the universe expand".

Alexander P. Yefremov
Prof., Dr. Sci., Director of Institute of Gravitation and Cosmology
of the Peoples' Friendship University of Russia

Preface to the English Edition

To mark the 100th anniversary of Einstein's general relativity the *Minkowski Institute Press* publishes the first English translation of a very rare book on general relativity (its only Russian publication was in 1924), which turned out to be the last book by A. A. Friedmann (co-authored with V. K. Frederiks). This is the first and the only published volume of a five-volume book project, brutally terminated by the untimely and tragic death of Friedmann on 16 September 1925.

In 1924 A. A. Friedmann and his colleague V. K. Frederiks began to publish a fundamental monograph on the theory of relativity. Unfortunately, only the first volume of the monograph was published in 1924 – V. K. Frederiks and A. A. Friedmann, Foundations of the Theory of Relativity, Volume 1: Tensor Calculus (Academia, Leningrad 1924) [В. К. Фредерикс, А. А. Фридман, Основы теории относительности, Выпуск I: Тензориальное исчисление (Академия, Ленинград 1924)].

Here is the description of the project by Frederiks and Friedmann themselves:

"Initially we intended to publish our book at once, but due to technical difficulties we divided it into five parts which will be published separately. The first volume of the book is devoted to the foundations of tensor calculus. The second will outline the foundations of the geometry of multi-dimensional spaces. The third will deal with the foundations of electrodynamics. Finally, the fourth and the fifth volume will be devoted to the foundations of the special and the general principle of relativity."

For a number of reasons the translation of this volume was not easy at all. The major reason was the introduction by Frederiks and Friedmann of notions and notations, which differ from those now used in tensor calculus. After several attempts we abandoned our initial intention to provide comments (as footnotes) because those numerous footnotes would rather distract the readers and prevent them from following Frederiks' and Friedmann's helpful and smooth introduction and development of the ideas of tensor calculus.

Another reason that made the translation difficult was the quality of the scanned pages of the book, which we used since we were able to find only the scanned version of the book. The scanned images were generally good, but in many cases it was virtually impossible to read the subscript and superscript of vectors and tensors and we had to infer them from the context. A third reason was the omission in the text of the numbers of some equations and we had to again rely on the context. A fourth reason was Friedmann's habit of writing very long sentences; our translation of his book *The World as Space and Time*, where we had to deal with his habit, was not of much help.

Despite the fact that this book was published in 1924 and despite the presence of some unconventional notions and notations in it, this is still a valuable book, because it is written by two deep thinkers, partricularly Friedmann who in 1922 had the deepest understanding of the cosmological implications of Einstein's general relativity when he first showed that the Universe may expand (which was later discovered by Hubble).

We would like specifically to thank Professor Alexander P. Yefremov who wrote the Foreword. We also thank Fritz Lewertoff who translated (on pp. 24-5) a passage from the original German publication of D. Hilbert's work *The Foundations of Physics* (*Die Grundlagen der Physik*).

Montreal, 21 December 2015   *Svetla and Vesselin Petkov*

CONTENTS

x

Preface

This book is the first volume of our work "Foundations of the Theory of Relativity." Our goal is to present the theory of relativity rigorously enough from a logical point of view, assuming that readers have knowledge of mathematics and theoretical physics not exceeding the level taught at the Russian universities and the technical universities.

Initially, we intended to publish our book at once but various technical problems forced us to divide it into five volumes. The first volume of our book is devoted to the presentation of the foundations of tensor calculus. In the second we will discuss the foundations of the geometry of multidimensional spaces. The third volume will be devoted to the foundations of electrodynamics, and, at the end, in the forth and fifth volume we will present the foundations of the special and general principle of relativity.

Specific technical difficulties of printing books containing complex mathematical language considerably delayed the publication of the first volume of our book. Two years passed from the moment of submitting the book till now. During this time, a number of works devoted to tensor calculus appeared in the foreign literature. In these works the notations were improved to some extent. We cannot, understandably, use these works in our book.

It is necessary to mention here, that two very good introductions to tensor calculus and partially to the geometry of multidimensional spaces have recently appeared. We have in mind fist of all the book J. A. Schouten, *Der-Ricci Kalkül*, Berlin, 1924, and also the book of H. Galbrun, *Introduction à la Théorie de la Relativité: Calcul Différentiel Absolu et Geométrie*, Paris, 1923.

In conclusion, we would like to thank the publisher "Academia" for doing everything possible to publish the first volume of our book with the necessary accuracy. Also we would like to thank E. P. Friedman for her invaluable help with the proofreading of the book.

1 August 1924

V. Frederiks
A. Friedmann

INTRODUCTION

Physics is not Geometry
and Geometry is not Physics.
(From conversations ...)

1. 1913 should be formally regarded as the year of creation of the general theory of relativity, because the first work of Einstein (coauthored with M. Grossmann) appeared then. It was called "Outline of a Generalized Theory of Relativity and of a Theory of Gravitation"[1]

But the first work where the fundamental ideas of Einstein were outlined with sufficient detail and clarity appeared in the proceedings of the Berlin Academy of Sciences on 19 November, 1914. And it is namely this work which had become a starting point, not only for Einstein's own research, but also for the work of a number of other great scientists – physicists, mathematicians and astronomers – such as Lorentz, de Sitter, Hilbert and others.

This work was called called "Formal Foundations of the General Theory of Relativity"[2]. This work attracted almost immediately great attention not only because it was a certain generalization of the theory of relativity which already made Einstein famous in 1906 and 1907, and which is now called "special" theory of relativity, but because of other reasons as well.

The new theory again and in its essence, touched upon an old question of physics on whose clarification, since the time of Newton, has not been reached any kind of success despite numerous attempts; a question which had stayed all the time aside from all other problems

---

[1]A. Einstein und M. Grossmann, Entwurf einer verallgemeinerter Relativitätstheorie und einer Theorie der Gravitation. B. Tuebner 1913.

[2]A. Einstein, Formale Grundlage der allgemeinen Relativitätstheorie, Preussische Akademie der Wissenschaften, Sitzungsberichte, 1914 (part 2), S. 1030-1085. With changes and additions it is printed in Annalen der Physik, 1916 Vol. LI p. 630 and after published in a separate issue by B. Tuebner.

of physics and which had not been connected in any way to the other physical phenomena. This was the question *of the nature of gravity*, whose clarification seemed impossible.

Recall that, on the one hand, Newton's discovery of the *law* of universal gravitation completely justified the words once said about him: "and God said: Let there be light and Newton came". On the other hand, for the same phenomenon of gravity La Rochefoucault wrote only several years after the death of Newton: "Gravity is a mystery of bodies, invented with the purpose to hide the shortcomings of our mind" (La gravité est un mystère du corps inventé pour cacher les défauts de notre esprit). Karl Neumann likens the action of gravity to two icy mountains, calling each other and moving to each other and wrote: "of course, such kind of a story should be regarded as belonging to the realm of *fables*; but we should not be laughing prematurely. Ideas, which seem to us so *strange*, lie on the foundations of the most advanced areas of the natural sciences and they are behind the glory of the most famous scientist."[3]

The above assertions reflect the two sides of the question which was again raised by Einstein: the foremost importance of the *law* of universal gravitation and the apparent incomprehensibility of its *nature*.

But the new theory, dealing with gravity, used completely different means beyond everything that had been regarded in physics as acceptable and having real meaning. The new theory applied in the purely experimental science, in relation to the notions of space and time, mathematical disciplines connected to systems of geometry, different from Euclidean geometry, and considered up to now as the highest degree of abstraction of all that exists.

At the end, the new theory, without making any special hypothesis, led apparently to the explanation of one, still unexplained, astronomical phenomenon, namely the very fast motion of the perihelion of Mercury, and predicted correctly another, earlier not observed but now found, phenomenon of deviation of a light ray in a sufficiently strong gravitational field.

No physical theory, as the general relativity principle, has been to such an extent a subject to extreme views – on the one hand, admiration and worship, and, on the other hand, of indignation and persecution, not only by physicists, astronomers, and mathematicians but also by people not competent in this field. For about slightly more than eight years of its existence the theory of relativity has become

---

[3]Our italics.

the subject of not only huge scientific, philosophical and popular literature, but also of numerous articles in magazines and newspapers, very often of quite low quality, creating for this theory a purely artificial success – some kind of fashion. As a result, the principle of relativity has been demonstrated even in cinematography and almost everybody thinks that he or she has the right to judge almost the most complex and difficult questions of physics. By the way, the last circumstance would not have been any trouble by itself, if it had not given rise to passions, even political passions, as for example in Germany, where Einstein had been subjected to different kinds of intimidation and threats, which could have hardly helped carrying out normal scientific work and adequate evaluation of his theory.

It is interesting to mention the opinions of some well-known physicists and mathematicians about this so exceptionally successful theory; some are very positive, others completely denying the importance of the theory of relativity.

J. J. Thomson said: "This is the most important result obtained in connection with theory of gravitation since Newton's days. Einstein reasoning is the result of one of the highest achievements of the human thought."[4]

M. Planck writes: "This is much more complicated compared to everything that has been created till now in the field of natural sciences and even in the realm of philosophy. The revolution in the physical concepts about the world could be compared by its deepness and volume only to the revolution made in its time by the Copernican world system."[5]

Other opinions have more modest character. W. Wien writes: "The theory of relativity is nothing more than a mathematical system of theoretical physics, which conclusions should be experimentally checked"[6]; "Many objections of theoretical and epistemological character do not touch on this theory as a physical theory ... not only the theory itself but also its consequences are not yet supported by the experiment."[7]

P. Lenard, pointing out the doubtfulness of the experimental confirmation of the theory, finishes his article against the theory of relativity (where he separates the issues of gravity and relativity) with the fol-

---

[4]Proceedings of the Royal Society. Meeting held on 6 November 1919 to discuss the results of the Eclipse expedition.

[5]Max Planck, Acht Vorlsungen über theoretishe Physik (Leipzig 1910 p. 117).

[6]W. Wien, Die Relativitätstheorie vom Standpunkte der Physik und Erkenntnislehre, Vortrag, Leipzig, Verlag von Johann Ambrosius Barth, 1921, p. 25.

[7]Ibid.

6

lowing words: "Further development of the theory will be clarified in the future. Only then it will become clear to what extent the gravitational principle – together with the already refuted even by everyday experiments general principle of relativity – will keep for itself at least heuristic meaning."[8]

H. Fricke comments rather rudely: "..., that the theoretician Einstein completely misunderstood the results of experimental physics."[9] Only this kind of thoughts could create interest in the theory of Einstein.

It is clear, that different scientists had favored and favour different sides of the theory, depending of what they consider in it the most important or interesting. This could be seen, by the way, from the titles, which have been given to it.

So H. A. Lorentz calls it "theory of gravitation"[10]; H. Weyl gives the title of a book, devoted to it, "Space, time, matter"[11]; D. Hilbert calls his article "Foundations of physics",[12] etc.

It is true that "relativity" and the phenomenon of gravity are studied in introductory physics, and it is also true that the subjects of study at this level are first of all questions concerning the applications of geometrical notions, the notions of time, force, mass, etc. It is necessary to have clear understanding of all these basic notions of physics and to be able to apply them – only then one could understand and appreciate the real value of the theory of Einstein. In our introduction we will only try, in general, to shed some light on the basic notions of physics. We will touch on them in the order in which they are usually discussed; first of all we will consider the notions of space and time, and after that the physical notions – mass and force (particularly the force of gravity). Of course, we will discuss them only in such detail which will be necessary for the understanding of Einstein's theory.

2. In Classical Mechanics and Physics, i.e. in Mechanics and Physics in the form, in which they were generally accepted before Einstein, and considered the only possible – in their ordinary presentations, the questions of the geometry of space and time were discussed very little. It is explained by the fact that the notion of space is taken simply from Euclidean geometry, which is well known, whereas the

---

[8]P. Lenard, Über Relativitätsprinzip, Aether, Gravitation; Leipzig. S. Hirzel. 1921. p. 44.

[9]T. Phil. H. Fricke, Regierungsrat; bei Hikner, Wolfenbüttel 1920, p. 5.

[10]H. A. Lorentz, On Einstein's theory of gravitation. Part III and IV Koninklijke Akademie van Wetenschappen te Amsterdam. Proceedings. Vol XX. VI. p. 2 1916.

[11]H. Weyl, Raum, Zeit, Materie. 4 Auflage, Berlin, Springer 1921.

[12]D. Hilbert, Die Grundlagen der Physik, Nachr. d. Ges. d. Wissensch. zu Göttingen 1915, 1917.

notion of time, existing in every person, differs only by the method of its measurement. In these relatively rare case when the authors of books, monographs, etc., on Mechanics or Physics wanted to be more precise and complete in their introductions, we find some discussions of space, time and their properties. Various authors do this, of course, differently where greater influence comes (to a various extent) from the corresponding epoch or from their personal views.

In Newton's "Philosophiae naturalis principia mathematica" we find the following assertions: "absolute space, by its nature, without relation to anything external, remains always similar and motionless" and "absolute, true, and mathematical time, in itself and by its nature without relation to anything external, flows uniformly."[13] Further, Euclidean geometry is applied to space followed by different kinds of explanations of measuring space quantities and time, and of the relative and absolute meaning of the obtained results. It is not necessary to criticize the essence of this approach; it is mentioned here simply as an example of a certain manner of dealing with the issues of space and time.

If we turn to more recent books, we find in some respects almost the same approach, although the words "absolute" space and time are avoided. In C. Schaefer's book "Einführung in die Theoretische Physik", 1919 (only the first two volumes appeared) it is written about space: "The properties of space are known from Euclidean geometry" ("Die Eigenschaften des Raumes sind aus der Euclidischen Geometrie bekannt", S. 2) and this is all that Schaefer said about space. With respect to time, he is more careful and devotes to it several pages (S. 2–4). Probably such a method of presentation was adopted by the author not to exhaust this question, but because of his unwillingness, caused by different considerations, to touch on it at all, or only on this specific point. But such an approach to the issues of space and time entirely corresponds to the accepted views on the Euclidean geometry, space and time, and do not take into account at all an important point, which did not exist in Newton's time but exists now, and which refers to the nature of the axioms of geometry particularly to their logical and experimental truthfulness. We have to give here some clarification on this question which we consider to be extremely important.[14]

As we know, initially the Euclidean geometry had developed through experimental means. Also, it is clear that the truthfulness of its basic

---

[13]I. Newton, Philosophiae naturalis principia mathematica (1687).

[14]In Poincaré's book *La Science et l'Hypothèse* all this is presented very clearly. We will repeat some of his explanations since they at present are very often forgotten.

8

assumptions had not been questioned at the time of Euclid, if not even earlier, regardless of any experiments. The Euclidean geometry had as if two kinds of truthfulness: one purely mathematical, absolute, and the other, experimental, relative, where this relativity depended only on the unavoidable inaccuracy of the physical measuring tools.

But the created by Gauss, Lobachevsky and Bolyai the first non-Euclidean geometries, and after that all other geometries different from the Euclidean, posed the question about the truthfulness of the axioms of geometry from a completely different perspective. It is known that the geometry of Lobachevsky[15] arose from the attempts to deduce the axiom of parallels from other axioms of Euclid and is at the same time: 1) an attempt to prove[16] the impossibility to carry our such a deduction, and 2) an attempt to prove the possibility of the existence of other geometries, different from the Euclidean geometry. Here we could not touch on the geometrical aspect of the question. For us, it is only important that every geometry, Euclidean or non-Euclidean, is constructed on the basis of certain axioms, and leads to logically correct consequences, and in such a way gives rise to the question of what is the experimental truthfulness of the Euclidean geometry, which forms the foundations of classical physics and mechanics. The clarification of this question is especially important, because in Einstein's theory we deal, unlike in the old mechanics, with geometries different from the Euclidean.

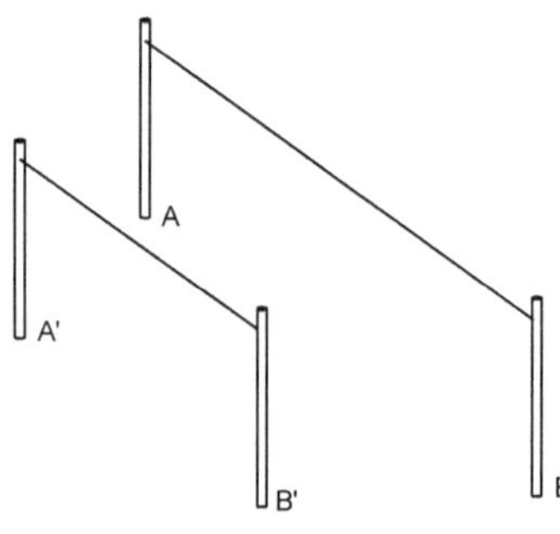

Figure 1

Take two pegs A and B (see Fig. 1), insert them into the ground and connect them with a thread; next take two other pegs A' and B',

---

[15] A book with three works by Lobachevsky on non-Euclidean geometry – N. I. Lobachevsky, *The Foundations of Geometry: Works on Non-Euclidean Geometry* – will be published by the *Minkowski Institute Press*, Montreal in 2016.

[16] The rigorous proof is to demonstrate the independence of the axioms of Euclidean geometry from each other; see §1, second chapter, the first volume of this book.

also insert them into the ground and connect them with a thread of the same length as that connecting A and B, but suppose that in the second case we used not the whole thread, but only part of it, say half of it. In such a way we present this purely physical operation as geometrical. Between the points A and B lies a segment of the straight line AB; between the points A' and B' lies the segment $A'B'$; the segment AB is twice as long as the segment $A'B'$. In such a way, the pegs are regarded as points, the thread – as a segment of a straight line. We can say: the thread "physically" measures the "geometrical length" of segments and thus we establish the link between physical phenomena and geometrical notions.

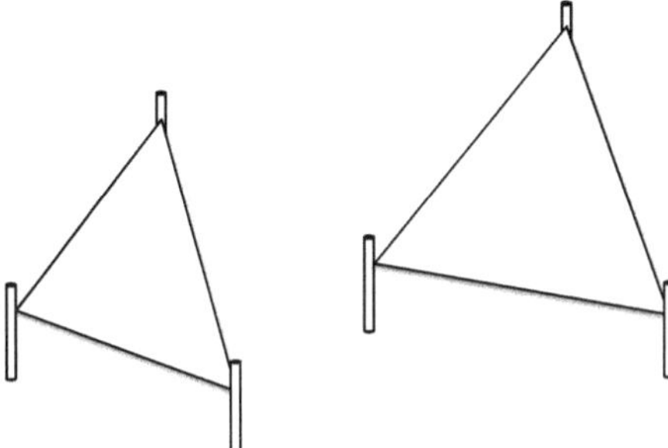

Figure 2

Take further three pegs A, B, C and three pegs A', B', C' (see Fig. 2), connect them with threads and also with some textile, which we cut in order that it fits the space between the threads connecting the pegs. Suppose that we need twice as much textile for the pegs A, B, C than for the pegs A', B', C'. Regard the pegs as points and the threads as segments of straight lines. The relative lengths of the threads give the relative lengths of the sides of the triangles ABC and $A'B'C'$. By using the known theorem of Euclidean geometry about the area of a triangle, determined by its sides, we find the relative areas of the triangles ABC and $A'B'C'$. Classical Mechanics, presupposing the correctness of Euclidean geometry, wants to say in this example that the ratio of the areas of triangles ABC and $A'B'C'$, calculated on the bases of the corresponding theorems, will be equal to two and that *it is not necessary* to compare the amount of textile stretched between the pegs. But it is evident that such a comparison was not necessary only until the discovery of the non-Euclidean geometries, in which the corresponding theorems of areas give other values for the ratio of the areas of the triangles ABC and $A'B'C'$. If we assume that not only in this experiment with textile, but in all other similar experiments, the predictions of Euclidean geometry are confirmed, then it would appear that the correctness of Euclidean geometry is proved experimentally. But such a proof is only apparent; in fact, it is based on two assumptions: first, on the coincidence of the calculated with the experimentally determined values, and, second, on regarding pegs as points, threats as segments of straight lines, stretch textile as part of a plane, etc. The coincidence of calculated and experimentally determined values will always lies in certain margins determined by the accuracy of our experiments: we can never be completely sure that there is no deviation from Euclidean geometry within those margins. The theory of relativity is an excellent confirmation of this. We will see later that non-Euclidean deviations in most cases are too small for our geodesic instruments – they will affect other experimental observations; therefore, the relative accuracy of observations does not allow us to make assertions of the absolute correctness of Euclidean (or any other) geometry.

But relative correctness also becomes meaningless, if we pose the question of whether we should always identify pegs with points, threats with straight lines, etc. This is most clearly evident from the experiment performed by Gauss to determine whether the sum of the angles of a triangle is equal or not to two straight angles. If the sum of the angles were equal to two straight angles, that would prove the axiom of parallels, and therefore the correctness of one of the axioms of Euclidean geometry. Three points – namely: 1) the peak Brocken of the Harz mountain range, 2) the peak Hohenhagen near Göttingen, and 3) the peak Inselsberg had been chosen by Gauss as the vertices of the triangle with a relatively large area of approximately 6000 km^2.[17] The angles had been measured with theodolite and the result of the measurements, within the limits of the observational accuracy, had been precisely what was expected on the basis of Euclidean geometry. What conclusions can be drawn from here? Obviously, in this experiment the geometrical triangle had been realized physically with the help of theodolites, regarded as points, and with light rays, regarded as the sides of the triangle. The physical phenomenon – a beam of light rays – is interpreted here as a straight line (similarly to the earlier example where a thread was interpreted as a straight line), and the assertion that the sum of the angles of the triangle is equal to two straight angles has meaning only because of the assumption that a beam of light propagates along a straight line. If the light rays are not regarded as straight lines, the proof of the sum of the angles in a triangle obviously fails. In such a way, Gauss' experiment presupposes

---

[17]J. C. F. Gauss, *Untersuchungen über Gegenstände der höheren Geodäsie*, Leipzig 1910.

that the propagation of light along straight lines was already *proved* in Physics. Such a proof can be demonstrated only if we can use another physical straight line, for example, assigning the properties of a straight line to a stretched thread. Then we can say that a light ray is as straight as a stretched thread. But if we suspect that the thread is not a perfect straight line (it may, for example, bend a little due to its own weight), we will find ourselves in an extremely difficult situation. It is true, that we can always use other physical instruments, consider other phenomena, but at some point we have to stop and say – it is this phenomenon and this instrument to which we *assign* the property of geometrical straightness; we will not have any longer any "physical" possibility to verify this assertion, regardless of any further suspicions on its correctness. Therefore, we see that the geometrical interpretation of a given phenomenon – for example, interpreting the thread as a straight line – is an *arbitrary* act. If the sum of the angles in the triangle in Gauss' experiment had not been equal to two straight angles, but to some other value, then we could regard the light ray as not a straight line and therefore that experiment could not prove anything. Only if we *decide* to interpret a light ray as a straight line (or any other physical phenomenon which, like a beam of light, can be physically regarded as a straight line), we are in a position to talk about correctness or incorrectness of the Euclidean or other geometries within the limits of accuracy of the corresponding observations. From experimental point of view, geometry is characterized by *conventional* truthfulness and *depends* on the *geometrical interpretation given to physical phenomena.*[18] Below we will see other example confirming this.

It follows from what was said above that *neither physics can be geometry nor geometry can be physics.* The type of geometry, employed in physics, will always depend on the chosen geometrical interpretation of physical phenomena. If we stick to the interpretation implicitly used in classical mechanics, the question of the correctness of Euclidean geometry from experimental point of view is completely meaningful. One of Einstein's contributions is that this question had been raised again and solved in a way different from that before him. It had been possible to imagine an interpretation different from that employed by classical mechanics – then the experiment would have given not that geometry which we have in the first case. In any case, *it is necessary to adopt one or another geometrical interpretation of physical phenom-*

---

[18]EDITOR'S NOTE: Here the reader's critical reading is especially necessary - Frederiks and Friedmann arrived at this conventionalist interpretation of geometry probably influenced by Poincaré.

12

ena; without it no physical theory is, of course, possible and we can assert that *physics without geometry does not exist*. But geometry exists in physics as long as we give geometrical interpretation to physical things. Through it and corresponding to it, a special geometrical space is created, which is called *physical*. Therefore, without some material objects, which can be interpreted geometrically, the physical space is unthinkable and empty physical space is simply nonsense.

Let us see how a geometrical interpretation is introduced in physics. One of the first tasks of physics is to define a method for finding the position of a given event or phenomenon. It is known that in geometry a position is specified with the help of coordinates. Each point in three-dimensional space is specified by three numbers which characterize this and only this point and therefore it can be distinguished from all other points. Cartesian (rectilinear), cylindrical, spherical, and many other coordinates can serve this purpose. Assigning three numbers to each point of the three-dimensional space will be called *arithmetization*. We stress that the arithmetization 1) is arbitrary, and 2) has nothing to do with the notion of measurement, not even with the notions of more or less (like, for example, the latitude and longitude, specifying a position on the Earth's surface, say nothing about distances).

Through a suitable geometrical interpretation the world of material things can be also arithmetized, which is as arbitrary as the arithmetization of the geometrical space. Such an arithmetization means that coordinates are introduced in the physical space. The transition from one coordinate system to another is a transition from one arithmetization to another.

By identifying the things, constituting space, and clarifying their relationships, we can divide the properties of things into two classes. Some things and properties will depend on the chosen (always arbitrarily) arithmetization of space, others will turn out to be unchangeable, no matter how we arithmetize space: the first things and properties will be called *non-intrinsic*, and the second – *intrinsic*. Let us compare the geometrical space with the already arithmetized physical space with the help of new geometrical interpretations. We will now find intrinsic and non-intrinsic properties of the physical space, depending and not depending on its arithmetization, respectively. For us, it is extremely important to identify the intrinsic things and their properties, because only with their help we can study space independently of its arbitrary arithmetization. We should, by the way, remark that studying its non-intrinsic things or properties also may be necessary; the best example is spherical astronomy.

The intrinsic properties of space can be expressed by assertions, whose form does not change in the transition from one arithmetization to another; a given transformation of coordinates corresponds to such a transition. This fact is expressed in the following way: the intrinsic properties are *invariant* under coordinate transformations.

As an example of such an invariant property we will give the length or the distance between two points. If we consider two points specified by two systems of numbers x_1, x_2, x_3 and x'_1, x'_2, x'_3, then we define a special function:

$$D\left(x_1, x_2, x_3; x'_1, x'_2, x'_3\right),$$

which should possess different kinds of properties (which will be discussed below). For two infinitely close points x_1, x_2, x_3 and $x_1 + \mathrm{d}x_1, x_2 + \mathrm{d}x_2, x_3 + \mathrm{d}x_3$ the function D obtains a certain expression, whose form is independent of the coordinate system. In this case the quantity D is called the distance $\mathrm{d}s$ between the two infinitely close points. For the three-dimensional space it has the following form:

$$\mathrm{d}s = \sqrt{g_{11}\,\mathrm{d}x_1^2 + g_{22}\,\mathrm{d}x_2^2 + g_{33}\,\mathrm{d}x_3^2 + 2\,g_{23}\,\mathrm{d}x_2\,\mathrm{d}x_3 + 2\,g_{31}\,\mathrm{d}x_3\,\mathrm{d}x_1 + 2\,g_{12}\,\mathrm{d}x_1\,\mathrm{d}x_2}$$

where $g_{11}, g_{22}, g_{33}, g_{23}, g_{31}, g_{12}$, are six functions of the coordinates x_1, x_2, x_3. In Euclidean geometry, in orthogonal and rectilinear coordinates, $\mathrm{d}s$ has the following form:

$$\mathrm{d}s = \sqrt{\mathrm{d}x_1^2 + \mathrm{d}x_2^2 + \mathrm{d}x_3^2}.$$

With respect to this formula, we will now only note that the invariant property of its expression is a property of foremost importance since it defines what is called the metric of space. This expression gives geometrical interpretation of those physical operations in the physical space, which we did with the threat stretched between the pegs, or with the light rays in Gauss' experiment.

We see that with the help of the fist stage of the geometrical interpretation the material world get arithmetized; different objects in it are interpreted as points with corresponding coordinates. The next stage of the geometrical interpretation identifies an invariant property of the already arithmetized material world – this is the distance between two points. Obviously, a whole sequence of other phases of the geometrical interpretation will follow, and all of them form what we call physics.

3. We can find a certain analogy between the notions of space and time. We just saw that some specific physical phenomena are given a

14

certain geometrical interpretation. Some *other* physical phenomena, which we call *motions*, need the introduction of the notion of *time*. We choose some motion arbitrarily and regard it as a special one. It is known that in classical mechanics this is the relative motion of the stars and the Earth. As the experiment shows, the position of a star relative to a given point on the Earth's surface changes periodically, and these periods are used to measure the universal time, which is the same for all space; we call this time "sidereal". All other motions are compared to this basic periodical motion and are interpreted as uniform or non-uniform with respect to it. *However, any other motion* – a falling body, the hands of a clock, a bouncing ball, etc. – can be regarded with equal logical rigour as a basic motion interpreted as a measure of time. Of course, the qualification of motion as uniform and non-uniform will be different from that in the case of sidereal time. Choosing sidereal time as basic, classical mechanics acted exactly as in geometry, where a light ray or a stretched thread were interpreted as straight lines. We will discuss neither the reason for this choice of time nor its significance in practical applications; it is only important to note here its *arbitrariness*.

Having chosen sidereal time, and studying other motions relative to it (that is, comparing them with the relative motion of the stars and the Earth), classical mechanics assumes that the very process of comparison is possible always and at any point of space, i.e. that time measured on the Earth's surface *can be also measured* at any point of space. The expression "can be also measured" means, of course, that we assume that there exist such physical apparatuses which make it possible to compare times at two different points. Clearly, the physical apparatuses should use some motion to be transported from point to point. If there existed motions with indefinitely great velocities, it would be simpler to find *physical* means to define simultaneity at different points. But contemporary physics imposes a limit on velocity, leading to difficulties in defining simultaneity, which the classical mechanics, allowing infinite velocities, did not encounter. Practically, as known, comparison of times (e.g. the readings of two clocks) is done visually, i.e., through light, which propagates with finite velocity. As the speed of light is very great, practically the transmission of clocks' readings with light signals can be regarded as instantaneous, but, of course, only within *certain* limits; going beyond these limits would constitute a big logical error, which can lead to conclusions contradicting experiment.

Let us consider setting time with a simple apparatus, which we will call a "light clock". This apparatus can serve to define time at a given point, and also, as we will see in the third volume of our work, it can be used as an apparatus with which we can synchronize distant clocks, that is, establish simultaneity at different points of space. Let point A (see Fig. 3) be the point at which we will record time. We place two mirrors at point A and at another point B. The two mirrors are parallel and their reflecting surfaces face each other. We place some source of light S between the mirrors. A light ray emitted at S will constantly travel from one mirror to the other due the reflection from them; recording the period of time between two successive reflections from A can serve as a method for defining time at A. The results obtained for the description of any motion with light clocks for different positions of B may not be proportional to one another, and also may not be proportional to the results obtained when sidereal time is used.

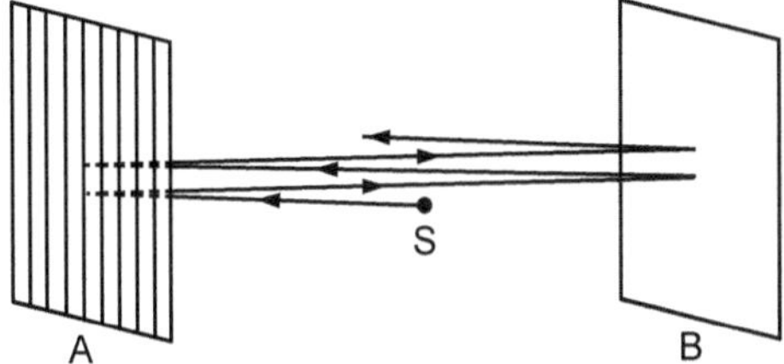

Figure 3

Above we arithmetized space by introducing three numbers which represent each of its points. We can arithmetize time in the same way by assigning a number to each of its moments. Any clocks, no matter how they run, provide a physical method for arithmetization of time. Any repetitive phenomenon can be used for this purpose. We have emphasized that the existence of invariant properties of space, independent of its arithmetization, and as a first example we gave the distance between two points, calculated by using their coordinates, which themselves were given by the arithmetization of space. It is easy to imagine the corresponding analogy in the case of time.

The arithmetized time will also possess invariant properties, independent of the method of its arithmetization. One of these properties, like distance, will be the interval of time between two moments: but the distance was a geometrical interpretation of a given physical phenomenon (recall the experiment with the thread and with light rays) – in a similar way, the interval of time should be also a geometrical interpretation of some physical phenomenon, for example, some motion. The choice of this motion is independent of the choice of the motion used for the arithmetization of time and is completely arbitrary. For

16

defining the "interval" of time we will use, from now on, the same light clocks, which we chose for its arithmetization. Therefore, their readings will serve not only as a method for counting the moments of time, but also for measuring the interval of time. If we count the moments of time by using some other clocks, for example, sidereal – which is always possible – then the difference between moments of sidereal time will no longer be (by our definition) an interval of time, exactly as the difference between the longitudes of two points on the equator does not represent their distance from each other.

The transition from one arithmetization of time to another means a transformation of time t to $\bar{t}$ by the substitution:

$$\bar{t} = \varphi(t).$$

The function $\varphi(t)$ may depend on the coordinates of the point, at which we carry out the transition from one arithmetization of time to another, so that we can write for the transformation of time the following equation:

$$\bar{t} = \omega_4(x, y, z, t),$$

where x, y, z are the coordinates of a given point.

If we change the arithmetization of space, going from coordinates x, y, z to coordinates $\bar{x}, \bar{y}, \bar{z}$, where for different moments of time the above transition of time is done by different methods, then the transformation formulas of time and space coordinates can be written in the following way:

$$\bar{x} = \omega_1(x, y, z, t),$$
$$\bar{y} = \omega_2(x, y, z, t),$$
$$\bar{z} = \omega_3(x, y, z, t),$$
$$\bar{t} = \omega_4(x, y, z, t).$$

Classical mechanics, treating time as being the same for all space, always viewed it as separate from the space coordinates. In classical mechanics space has three dimensions, and time – one. But we saw that points (of space) and moments (of time), distances between points and intervals between moments, are closely related notions; that the transition from one arithmetization of space and time to another is represented by the same common transformation formula. We already know from the special principle of relativity that the physical space and time are not as independent[19] from each other as it was thought

---

earlier. It seems natural, at least formally, to regard time as a *fourth* dimension, which is added to the three purely space dimensions. We will see later that the fourth dimension cannot be completely identical to the space dimensions due to the so called *principle of causality*, which does not allow that cause and effect (for example, birth and death) be interchanged in time as the order of two points located on a straight line.

The purpose of the above comments was to provide a general idea of the significance of geometry in physics as a geometrical interpretation of given physical phenomena and time, as a geometrical interpretation of other physical phenomena. Obtained in this way, geometry and time or, in the usual terminology, physical space and time, taken together, form a four-dimensional geometry.

The elements of this four-dimensional geometry represent certain physical phenomena. The natural question here is to what extent the four-dimensional geometry correspond to the experimental facts. We will see later what the experimental evidence will say on this question, but even now we can say that its definite answer is unthinkable without further study of the basic notions of physics, namely the notions of force and mass. We will now turn to the analysis of these notions.

4. The first purely physical notions, which are introduced in the first branch of physics – mechanics – are the notions of force and mass, and the basic laws of classical mechanics involve first of all force and motion of mass. These two notions are *basic and independent*, and experimental physics gives us means, in each physical case, to *assign* certain numerical values to these notions, i.e. to *measure* force and to *measure* mass. It has always been the goal of physics and mechanics to give these notions precise mathematical definitions. How difficult it is to achieve this goal can be seen from Newton's definition according to which *"the quantity" of matter* (mass) is the measure giving the proportionality of its density and its volume. "The *applied force* is the action on a body which changes its state of rest or its uniform motion."[20] It is clear that these definitions are insufficient as seen from the fact that during the last half a century, there appeared many works criticizing them. A great amount of effort had been spent on finding more precise and comprehensive definitions.

The forces which we encounter in classical physics are as we know of very different nature. We will divide them in two main groups. We

---

знаем уже из специального принципа относительности, что понятия пространство физического и времени не так *зависимы* друг от друга, как это думали раньше"), but I think it is a typo and the word should be "independent".

[20]I. Newton, Philosophiae naturalis principia mathematica (1687).

18

will put in the first group the so called inertial forces – the centrifugal
force, the Coriolis force, and so on. Mechanics regards these forces
as fictitious or apparent, because the motion which they as if explain
is "in fact" not caused by them. The second group contains all other
"real" forces, which we will divide into three categories: 1) the force
of gravity, 2) the electromagnetic force, and 3) all other forces such
as, for example, molecular forces or the forces of inter-atomic inter-
actions, elastic forces, friction forces, etc. In recent years, a view has
been emerging in physics that the forces of the third category are ultimately also electromagnetic in origin. Even if this view should not be
regarded as absolutely proved, it may be expected that the so-called
"electromagnetic picture" of the world – as far as the phenomena in-
volving forces of the third category are concerned – will triumph at
the end. Undoubtedly, there are very small chances that the inertial
force and the force of gravity will be reduced to electromagnetic forces,
despite some attempts (dealing with gravity[21]) in this direction.

The difference between the forces of inertia and gravity is the role
of mass in them. The force of universal gravitation f_g between two
point masses is expressed by the formula

$$f_g = k \frac{m\,m'}{r^2}$$

where k is the gravitational constant, m and m' are the masses of two
material points, and r is the distance between them. Here m and m'
are regarded as if active masses, inducing the forces of gravity between
the material points. The centrifugal force f_c is one of the inertial forces
in the first group of forces, discussed above:

$$f_c = \frac{m\,v^2}{r}$$

---

[21] See Ritz. Gesammelte Werke, Paris, Gauttier Villars, 1912. *La Gravitation*, p.
489. In this paper Ritz explains why the so-called electromagnetic theory of Zöllner
and Messotti and its modifications remain in their essence outside the framework
of the contemporary electromagnetic theories. It is true, in this paper Ritz outlines
a program which, in his view, every electromagnetic theory of gravitation should
satisfy, and in which, by the way, the gravitational constant can be defined through
purely electromagnetic phenomena (which is impossible in the theory of Zöllner and
Messotti). Ritz thinks that such a program can be realized. However, so far this
has not been done and the difficulties that should be overcome are so great that it
is unlikely that it can be done in the foreseeable future.

See also the paper I. Zenneck, Gravitation, Encycloped. der Math. Wissenschaften. Bn. V. 2 p. 25 and its concluding phrase (p. 69) ... "But in this way
in the beginning of the 20th century we returned again to the viewpoint of the
18th century – to the view that gravitation is regarded as a fundamental property
of all matter (Fundamentaleigenschaft aller Materie)."

where v is the velocity of the material point, r is its distance from the axis of rotation, m is its mass, which according to classical mechanics does not induce the force f_c; if this force is compensated by some other force, then the greater the mass m the lesser v and the lesser the normal acceleration of the particle. The role of mass appears to be passive. In the first case (in the formula for the force of gravity) the mass is called *gravitational*, whereas in the second case (in the formula for f_c) – *inertial*. Both masses are determined by experiment, which demonstrates that with great accuracy, provided by the most precise contemporary observations, they are numerically equal (up to 3×10^8 of their values). We should note that classical mechanics ignores the fact that no experiment found any difference between the two masses.

Classical mechanics regards the forces of inertia as fictitious, but according to it the force of gravity is an example of a *quite* real force. Before Einstein, there had been many more or less unsuccessful attempts to create a theory of this force, that is, to explain it through other forces or phenomena of Nature.[22] Einstein's theory, unlike the other theories of gravitation, regards the force of gravity as belonging to the first group of forces. This, first of all, explains why there is no difference in principle between inertial and gravitational masses; and also, if we accept that all other physical forces are electromagnetic in origin, then all forces in Nature will turn out to be either inertial or electromagnetic.

To get an idea of the potential of Einstein's bold theory we will have to turn to how the basic mechanical laws are formulated. This is done in classical mechanics in the following way: according to it our space is Euclidean, our time is the universal sidereal time; further classical mechanics establishes the link between the accelerated motion of a material point, the force causing this motion, and the mass of the accelerating point. According to this link the material point will move uniformly along a straight line – by inertia – as long as no force acts on it. First of all, Einstein's mechanics abandons the a priori assumption that the physical space is Euclidean, and that the physical time is the universal sidereal time. As indicated above, based on the astonishing coincidence of inertial and gravitational masses, Einstein's theory regards the force of gravity as an inertial force. This brilliantly original hypothesis makes it immediately possible to determine experimentally the non-Euclidean character of the physical space. Rough experiment appears to show that the geometry of the physical space is Euclidean (more precisely: pseudo-Euclidean geometry of the four-dimensional

---

[22]See Zenneck, 1.

space; see the second volume of this work). But within the experimental accuracy there may exist such small deviations from the Euclidean geometry of the four-dimensional space, which can explain motions, traditionally attributed to gravity, as motions by inertia. In the ordinary three-dimensional Euclidean space the rectilinear trajectory of a uniformly (i.e. inertially) moving material point can be regarded as a projection of the straight line representing the inertial motion of the point in the four-dimensional space (whose fourth coordinate is time); motion by inertia is represented in the four-dimensional space by a line which is the shortest distance between any two points on the line in this space, that is motion by inertia is represented by a straight line in the four-dimensional space. Two features of inertial motion – rectilinearity and uniformity – are combined in one feature in the four-dimensional space – the shortest distance (in a four-dimensional sense).

According to Einstein this single feature of inertial motion applies to material points, represented by lines in a non-Euclidean four-dimensional space, but the lines of the shortest distance in such a space are not straight lines in the ordinary sense. The trajectories of planets (relative to the Sun a planet can be regarded as a material point) will be, for example, projections of the lines of the shortest distance in the non-Euclidean four-dimensional space onto the three-dimensional space, and therefore the planets' motion turns out to be inertial in Einstein's theory. Conversely, the non-Euclidean character of the physical space can be determined from the motion of the planets, i.e. from experiment.[23]

As indicated above, the difference between the classical Euclidean space and the new non-Euclidean space is so small that it remains within the limits of error in the laboratory and geodesic observations. But in some exceptional cases this difference can be detected experimentally, and Einstein's results are immediately confirm by experiment, whereas classical mechanics has to make special hypotheses to explain those experimental observations. For example, the motion of the perihelion of Mercury and the deviation of a light ray by a massive body (the change of the apparent position of stars near the Sun's disk).

Let us discuss in more detail the formulation of the basic laws of mechanics in relation to the concept of "relativity."

As we have already indicated, the first step in the description of the phenomena of Nature is the arithmetization of the physical space,

---

[23]For the sake of brevity, the terminology here is not quite rigorous. For a detailed treatment of this question see the third volume of our work.

i.e., the introduction of coordinate systems for identifying the positions and times of the occurring events. The choice of coordinate systems is completely arbitrary. We use rectilinear, spherical, cylindrical, and many other coordinate systems, and fix their origins at some points of the physical space. Of course, it has been always clear that the occurring events do not depend on the arbitrary choice of the coordinate system. But we describe events not only by determining their positions and the times when they happen. Through a suitable geometrical interpretation we *assign* extension in space and duration in time to matter, we *measure* forces, masses, etc., and after that we determine numerical relations between these quantities, which we call *laws of Nature*. The basic numerical relations – the laws of Nature – formulated by classical mechanics generally depend on the chosen coordinate system, but there exist such coordinate systems, which have exceptional status with respect to these laws. These exceptional coordinate systems are equivalent. They posses the property that they can be in relative rectilinear and uniform motion (in Euclidean sense) and are called *inertial* coordinate systems. This is the meaning of "relativity" in classical mechanics and the motion by inertia is closely related to it. If a body is at rest in an inertial system, it moves rectilinearly and uniformly in another inertial system. But in the first system the body is not subject to any forces (otherwise it would not only move, but move with an acceleration). Therefore the same holds in the second inertial system as well.[24] That is why we say that, with respect to the second inertial system, *the body moves rectilinearly and uniformly by inertia.* If there is a gravitational force, it should change the rectilinear and uniform motion by inertia in an inertial system in classical mechanics. Here we will leave aside the questions of the practical significance and of the real existence of inertial systems, and also of the difficulties, general or others, related to them. These questions provoked many debates and many works were devoted to them. Undoubtedly, the germ of Einstein's theory can be found in these works.

If, according to Einstein's theory, motion due to gravity and inertial motion are of the same nature, then it is clear that not only the inertial systems of classical mechanics should be equivalent. If this was really so, then inertial motion could be only rectilinear since motion due to gravity is not rectilinear (in Euclidean sense) and is not uniform. According the Einstein, inertial systems (which are equivalent) can move relative to one another arbitrarily, and then, through the principle of inertia, one can explain any motion, which classical

---

[24]EDITOR'S NOTE: As required by the equivalence of the inertial systems.

22

mechanics attributes to gravity. Instead of the restricted relativity of classical mechanics, Einstein formulated general relativity: *the fundamental laws of Nature do not depend on the coordinate system*; they can be presented in a form independent of any coordinate system. Mass and force manifest themselves in a new light: masses are caused not by the force of gravity, they only induce the geometry of the physical space. Motion, attributed to gravitation, is inertial motion in the non-Euclidean physical space.

The relativity of classical mechanics finds its mathematical expression in the fact that its fundamental laws do not change their form when the coordinates of one inertial system K are replaced by the coordinates of another inertial system $\overline{K}$. If x, y, z and t are the coordinates and time in K, $\overline{x}, \overline{y}, \overline{z}$ and $\overline{t}$ are the coordinates and time in $\overline{K}$, then the transition form K to $\overline{K}$ in classical mechanics is given by the transformation:

$$\overline{x} = x - v_x t, \quad \overline{y} = y - v_y t, \quad \overline{z} = z - v_z t, \quad \overline{t} = t,$$

where x_x, v_y, v_z are the three components of the constant velocity of $\overline{K}$ relative to K. The laws of mechanics are expressed equally in both variables $\overline{x}, \overline{y}, \overline{z}, t$ and $\overline{x}, \overline{y}, \overline{z}, \overline{t}$: time t, as discussed above, is universal in classical mechanics and is therefore the same in all coordinate systems.

In Einstein's general relativity, any two coordinate systems K and $\overline{K}$ can be related by any arbitrary relations:

$$\overline{x} = \omega_1 (x, y, z, t),$$
$$\overline{y} = \omega_2 (x, y, z, t),$$
$$\overline{z} = \omega_3 (x, y, z, t),$$
$$\overline{t} = \omega_4 (x, y, z, t).$$

All such systems are equivalent; in all of them the laws of mechanics have the same mathematical expression independent of the functions $\omega_1 \omega_2 \omega_3 \omega_4$ The time $\overline{t}$ may not be equal to t; in different systems and in different points of a given system, time may be defined and measured differently.

Tensor calculus, the theory of invariants, differential geometry, and calculus of variations gave Einstein the necessary mathematical means for finding these laws of mechanics. The laws formulated by Einstein take the place of Newton's three fundamental laws and his immortal law of universal gravitation. The fact that Einstein's theory employs a complex mathematical apparatus is often used to criticize the theory;

it is argued that due to its complexity it cannot replace the "simple" theory of Newton and classical mechanics. Regarding this we will say the following. Sometimes simplicity may be apparent, purely external, as, for example, the inertial systems of classical mechanics. Moreover, Einstein's theory does not exclude classical mechanics at all; on the contrary, classical mechanics is contained in general relativity as a first approximation. The general principle of relativity does not eliminate the special one; it only determines the domain of applicability of the special principle of relativity. In the limits of a physical laboratory it remains valid as before. Historically and logically, the special principle of relativity is the link between classical mechanics and Einstein's theory. Einstein and Newton do not contradict each other, but Einstein goes beyond Newton.

5. In the conclusion of the present introduction we will consider the question, which is undoubtedly of vital interest to those undertaking the difficult task of mastering the ideas and formalism of the general principle of relativity – what is the real meaning of this grandiose concept, is it worth spending so much effort for overcoming the logical and technical difficulties encountered by anyone who embarks on conquering the domain of thought introduced by the general principle of relativity?

Three sides characterize the significance of the general principle of relativity. First of all, this concept provides a powerful impetus to the *axiomatization of physics*. Further, the general principle of relativity provides an amazing unity of the physical worldview. Finally, this theory, supported by astronomy, makes it possible to reveal the secrets of the *macroscale of the universe*. Let us consider these three sides of the general principle of relativity in detail.

Roughly speaking there are three periods in the evolution of human thought. In the first the most powerful period, man through experiment accumulates a huge amount of facts; in mathematics this period is best represented by the richness of facts which had been accumulated by humankind at the dawn of the scientific advancement prior to the appearance of the Hellenic culture. In the second period the huge cultural heritage from the fist period is logically processed and schematized and as a result *science* is created; this period in mathematics is characterized for example by the epoch from Euclid and Pythagoras till XIX century. Finally, the third period is a period of introducing the *axiomatization of knowledge*, a period, which can be characterized as a time of senile skepticism. In this period a well defined difference between the logical content of science and its material interpretation appeared; the logical content of science is presented as a kind of a well

24

formulated system of uncontroversial and independent axioms; at the end in the same period other systems of axioms are created and an "imaginary" science is constructed, which seemed logically perfect, but is more difficult to interpret from material point of view. The period of skepticism in mathematics started in the beginning of of XIX century with (Gauss, Lobachevsky, Bolyai) and continued to take shape at the end of XIX and the beginning of XX century (Lie, Klein, Hilbert).

The general principle of relativity is obviously the first concepts in physics, containing the sign of the transition to the third period of axiomatic skepticism . The colossal amount of experimental results, the remarkable use of the mathematical apparatus, and as a result, emergence of a large number of deep and witty hypothesis and theories, brought physics to a great advancement at the end of the past and the beginning of today's XX century. But regardless of all the success of physics during that period the logical part of the hypothesis and theories had not been clear enough and unsettled (quantum theory, electronic theory, the theory of matter and etc.).

From here the need for a logical rework of the huge scientific content of physics emerges; the axiomatization of mathematics starts to push physics in this direction. The appearance of general principle of relativity uniting theoretical physics and "imaginary" geometry, and asking directly the question about space and time, obviously represented a powerful push to the logical development of contemporary physics in the spirit of its axiomatization. For now such axiomatization is still not present, but the path toward it is already cleared up and the eyes of a genius, seeing farther and better than other people, feel the possibility of creating compatible and independent axioms, from which all physics will follow with the same logical flow the as geometry or arithmetics follows from a non-contradicting system of its axioms. The article "The Foundations of Physics" ("Die Grundlagen der Physik," 1 Mittelung) by David Hilbert ends with the following words: "we see that not only our ideas about space, time, and motion change according to Einstein's theory, but I am also convinced that its basic equations will make it possible to elucidate the innermost and until now hidden processes inside the atom and, most importantly, it will become possible to relate all physical constants with mathematical constants, and this, in turn, shows that it will become possibles in principle to transform physics into a science similar to geometry; this will be a triumph of the axiomatic method which in the discussed here questions makes use of the mighty instruments of analysis, namely, variational calculus and the theory of invariants."[25]

Hence, the general principle of relativity starts special research leading to axiomatization of physics. In physics, even more than in geometry, side by side with axiomatization of science itself stands the question of the interpretation of its material world; dealing closer with the material world than the geometry or arithmetics physics will need much more attention while interpreting its notions of things belonging to the material world. This situation should create conditions for axiomatization of physics, completely different from the same process in geometry and arithmetics.

It is clear that as soon as the system of physical axioms will be formed,a tendency will appear (due to the need to prove the independent axioms of this system) to develop science, based on a system of axioms that is different from that, which will be defined by us and which obviously will correspond to the most natural interpretation. In other words, together side by side with the "natural" physics it arises a system of "imaginary" physics. These systems will find their interpretations in the material world like the non Euclidean geometry was interpreted by using elements of the ordinary Euclidean geometry. The creation of imaginary physics should not be considered fruitless work, because such a construction, besides its purely logical significance, gives the possibility to consider many properties of "natural" physics, the discovering of which by its complexity, could have been extremely difficult. Of course all of that work belongs to the future and physics should go a long way to reach such logical perfection.

The second aspect, shining in the general principle of relativity - is the aspect of creation of united physical worldview. The need for such unity is proved by the whole historical development of physics in the last decade. We already discussed the numerous theories and hypothesis, which appeared and are appearing now in physics; they should be united and also should be given the possibility to be viewed from united point of view. And the principle of relativity goes in this direction, sometimes with uneven, but always with ingenious steps. Already Einstein, connecting physics and geometry provided the possibility to explain the inexplicable universal gravitation; this path is followed further by Weyl. Creating new spaces he connects geometry, gravitation and matter; in his views all manifestation of gravitation or of electromagnetic processes are only aspects of the properties of the four-dimensional world. The interpretation of Weyl's ideas could not yet be considered experimentally proven, but the grandiosity of that

---

[25]Translated from the German by Fritz Lewertoff.

concept makes us approach it very carefully.

If the axiomatization of physics belongs first of all to the field of logical thought, if the unity of the physical worldview is essential for the physicist-philosopher, then the third aspect of the principle of relativity should most of all be of interest for the physicist-experimenter. The principle of relativity makes it possible to arrive at ideas about the macroscale of the Universe; these ideas are unexpected and could be checked constantly by experiment. It is true that these ideas are very often poor and not well justified due to the insufficient advancement of astronomy and lack of solid experimental means but nevertheless even the present knowledge which they give about our Universe shows the genius of the creator of principle of relativity.

Up to now we have been considering only the importance of the principle of relativity for physics; this theory had played huge role although indirect, in the development of such mathematical disciplines as tensor calculus, differential geometry and the theory of invariants. Hilbert and Weyl are the leading mathematicians who use the impetus of the theory of relativity to develop and improve the above mentioned mathematical disciplines.

We, fortunately, cannot see the future and we do not know whether the epoch of axiomatization, epoch of skepticism, are final hours of knowledge... But even if it were like that, the logical beauty of the end would be sufficient to welcome the principle of relativity.

1   The notion of tensors

1. Vector calculus is an extremely useful tool for studying and solving many problems of geometry, mechanics and physics, when for description phenomena we use, on the one hand, three-dimensional space, contrasting space to the idea of time, and, on the other hand, choosing rectangular coordinate system. The transition to curvilinear coordinates for the description of phenomena of the external world and to the four-dimensional "world", where time does not differ much from coordinates, which forces us to change and generalize the vector calculus, replacing the specific kind of affine (coinciding with the ordinary) vectors with vectors and tensors of a general kind.

Before discussing the notion of tensors of the general kind, we will discuss some well known questions of vector calculus. Usually, by *vector* we mean a notion which is characterized by three quantities (the components of the vector on the coordinate axes). Such kind of a formal definition of vectors, is convenient for many questions, but does not say too much about the physical significance of vectors. Almost all vectors in mechanics and physics, serving to describe the phenomena of the external world, possess properties, which do not change when we change one rectangular coordinate system with any other rectangular coordinate system. In fact, for example, the vector of force acting upon a point will not change as soon as we go from one coordinate system to another. So, the idea of the vector having a given physical significance should be linked with certain independence of this vector from the chosen rectangular system. In accordance with this, the components of a vector in a coordinate system should, in a given way, depend on the components of the same vector in another coordinate system.

Let the first coordinate system be $(O\,x_1\,x_2\,x_3)$ (see Fig. 4) and let the second coordinate system be $(\overline{O}\,\overline{x}_1\overline{x}_2\overline{x}_3)$. Let α_1, α_2, α_3 be the coordinates of O relative to the first coordinate system, and let α_l^i be

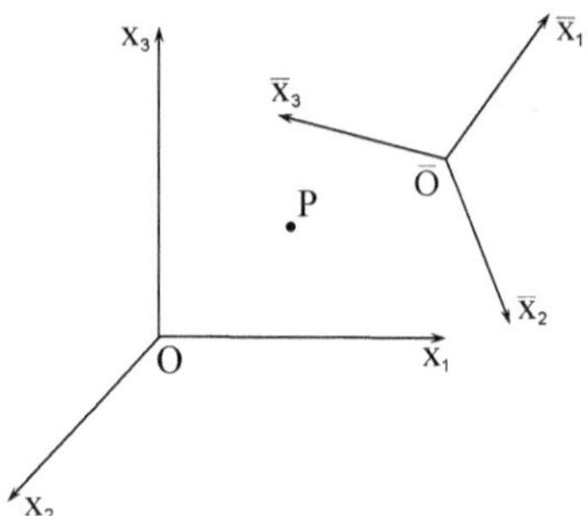

Figure 4

the cosine of the angle between the axis $\overline{Ox_i}$ and the axis Ox_l. We assume that the coordinates of point P in the first system will be x_i and in the second $-\ \overline{x}_i$ $(i = 1, 2, 3)$. Then we will have the following formulas for the transition from one coordinates to other coordinates:

$$\overline{x}_i = \alpha_i + \sum_{l=1}^{3} \alpha_l^i x_l, \quad (i = 1, 2, 3), \tag{1}$$

where the quantities α_l^i could be written in a table of the cosines of the angles between the coordinate axes

	$\overline{x}_1$	$\overline{x}_2$	$\overline{x}_3$
x_1	α_1^1	α_1^2	α_1^3
x_2	α_2^1	α_2^2	α_2^3
x_3	α_3^1	α_3^2	α_3^3

There exist six well known relations between the quantities α_l^i which could be written in the following form:

$$\sum_{i=1}^{3} \alpha_r^i \alpha_s^i = \delta_r^s, \quad (r, s = 1, 2, 3,) \tag{2}$$

where the symbol δ_r^s denotes $+1$ when $r = s$ and 0 when $r \neq s$.

Using these relations we can solve the equation (1) for x_i and write the following formulas:

$$x_i = \sum_{l=1}^{s} \alpha_i^l (\overline{x}_l - \alpha_l) \tag{3}$$

We will denote vectors with **bold** Gothics letters, whereas their components along the axes x_1, x_2, x_3 will be denoted by the corresponding normal (not **bold**) Gothics letters with subscripts $1, 2, 3$; in this way, $\mathfrak{a}$ will be a vector with components $\mathfrak{a}_1, \mathfrak{a}_2, \mathfrak{a}_3$. Generally speaking, when we go from one coordinate system to another, we will have to find out in what the vector $\mathfrak{a}$ will be transformed, without considering its physical meaning. We can assume that as a result of the coordinate transformations the vector $\mathfrak{a}$ will transform to another vector $\bar{\mathfrak{a}}$ with components (already along the new coordinate axes): $\bar{\mathfrak{a}}_1, \bar{\mathfrak{a}}_2, \bar{\mathfrak{a}}_3$. If we define the vector only as a set of three quantities, then we are completely free to choose the vector $\bar{\mathfrak{a}}$ in which the vector $\mathfrak{a}$ will be transformed as a result of the coordinate transformation. In most physical situations we deal with vectors which transform, as a result of coordinate transformations, by formulas which are analogous to the formulas by which the coordinate points change in the transition from one coordinate system to another:

$$\bar{\mathfrak{a}}_i = \sum_{i=1}^{3} \alpha_l^i \mathfrak{a}_i,$$

$$(i = 1, 2, 3). \tag{4}$$

$$\mathfrak{a}_i = \sum_{i=1}^{3} \alpha_i^l \bar{\mathfrak{a}}_l$$

Equations (4) correspond to the following geometrical property of vectors: if a vector, as usual, is represented by an arrow coming from point P (with its beginning at point P) and having components $\mathfrak{a}_1, \mathfrak{a}_2, \mathfrak{a}_3$ along the coordinate axes of the first system, this arrow will not change during the transition to the second coordinate system and its components along the axes of the second system will be $\bar{\mathfrak{a}}_1, \bar{\mathfrak{a}}_2, \bar{\mathfrak{a}}_3$, which are obtained from $\mathfrak{a}_i$, with the help of formula (4).

In such a way, we can define a vector in the following way:

If for each of the possible rectangular coordinate systems we have a set of three quantities, transforming under coordinate transformations by the formulas (4), then such a set of three quantities defines for each of the possible rectangular coordinate systems the notion of a vector.

It is not difficult to give several examples illustrating this notion of a vector. The vector radius $\mathfrak{r}$ of the point $P(x_1, x_2, x_3)$ relative to point $P_0(x_1^{(0)}, x_2^{(0)}, x_3^{(0)})$; $\mathfrak{r}_1 = x_1 - x_1^{(0)}$, $\mathfrak{r}_2 = x_2 - x_2^{(0)}$, $\mathfrak{r}_3 = x_3 - x_3^{(0)}$ is a vector.

The velocity $\mathfrak{b} = \frac{dx}{dt}$ and the acceleration $\mathfrak{a} = \frac{dx}{dt}$ are vectors; here t denotes time.

The vector gradient of a scalar is the vector:

$$\mathfrak{G} = \operatorname{grad} f, \quad \mathfrak{G}_1 = \frac{\partial f}{\partial x_1}, \quad \mathfrak{G}_2 = \frac{\partial f}{\partial x_2}, \quad \mathfrak{G}_3 = \frac{\partial f}{\partial x_3},$$

where f is a given function of the coordinates.

It is clear from this example, that for the vector gradient for any coordinate system there exists a law of assembling of its components; in order to clarify whether the vector is a gradient of a physical vector or not, it is necessary to check whether its components in different coordinate systems satisfy the condition (4) or not? If $\overline{f}$ is the function of $\overline{x}_1$ in which f is transformed when the system of coordinates is changed, then the components of $\overline{\mathfrak{G}}$ in the new system of coordinates will be:

$$\overline{\mathfrak{G}}_i = \frac{\partial \overline{f}}{\partial \overline{x}_i}, \quad (i = 1, 2, 3,),$$

but:

$$\frac{\partial \overline{f}}{\partial \overline{x}_i} = \sum_{l=1}^{3} \frac{\partial f}{\partial x_l} \frac{\partial x_l}{\partial \overline{x}_i} = \sum_{l=1}^{3} \frac{\partial f}{\partial x_l} \alpha_l^i.$$

In such a way:

$$\overline{\mathfrak{G}}_i = \sum_{l=1}^{3} \alpha_1^i \, \mathfrak{G}_1,$$

i.e. the vector gradient is really a vector.

It is not difficult to find a number of examples, giving tree numbers, which do not define a vector in the above sense.

For example the components:

$$\sigma_1 = \frac{\partial a_2}{\partial x_2} + \frac{\partial a_2}{\partial x_2}, \quad \sigma_2 = \frac{\partial a_2}{\partial x_3} + \frac{\partial a_2}{\partial x_1}, \quad \sigma_3 = \frac{\partial a_2}{\partial x_1} + \frac{\partial a_1}{\partial x_2},$$

playing some role in the theory of elasticity (compare shifts which are parallel to the coordinate planes), do not define a vector, which is not difficult to check by reversing the direction of the axis x_1 and leaving the other axes unchanged:

$$\overline{x}_1 = -x_1, \ \overline{x}_2 = x_2, \ \overline{x}_3 = x_3, \ \overline{a}_1 = -a_1, \ \overline{a}_2 = a_2, \ \overline{a}_3 = a_3,$$

$$\overline{\sigma}_1 = \sigma_1, \ \overline{\sigma}_2 = -\sigma_2, \ \overline{\sigma}_3 = -\sigma_3,$$

in such a way σ_1, σ_2, σ_3 are not components of a vector.

It is important to notice, that many of the so-called "vectors", serving for describing the external world, are not what which we called vectors. For example, the vector product of two vectors $\mathfrak{U}$ and $\mathfrak{B}$ denoted by $[\mathfrak{U}, \mathfrak{B}]$ and defined in every coordinate system by the following components:

$$\mathfrak{G} = [\mathfrak{U}, \mathfrak{B}].$$

$$\mathfrak{G}_1 = \mathfrak{U}_2\mathfrak{B}_3 - \mathfrak{U}_3\mathfrak{B}_2, \quad \mathfrak{G}_2 = \mathfrak{U}_3\mathfrak{B}_1 - \mathfrak{U}_1\mathfrak{B}_3, \quad \mathfrak{G}_3 = \mathfrak{U}_1\mathfrak{B}_2 - \mathfrak{U}_2\mathfrak{B}_1,$$

do not define a vector; in fact, reversing the direction of the axis x_1, and leaving the axes x_2, x_3 unchanged we find:

$$\overline{\mathfrak{G}}_1 = \mathfrak{G}_1, \quad \overline{\mathfrak{G}}_2 = -\mathfrak{G}_2, \quad \overline{\mathfrak{G}}_3 = -\mathfrak{G}_3,$$

which equations show, that $\mathfrak{G} = [\mathfrak{U}, \mathfrak{B}]$ is not a vector in the sense mentioned above.

It is well known, that each transformation of rectangular coordinate systems could be done through three operations:

1) translation of the origin of coordinates, 2) rotation of the coordinates axes, as a solid body, and 3) reflection similar to reflection in a mirror in one of the coordinates planes, which correspond to a reversal of the direction of one of the axes, while the rest of the axes do not change their direction. The first two operations are called *coordinate transformations with the help of motion*, the last - *coordinate transformations with the help of reflection.*

The vectors, which we defined, possess the properties to transform their components by the formula (4), in other words, *do not change* under coordinate transformations with the help of motion as well as with the help of reflection. They are called *polar vectors.*

They exist however a class of the so called "vectors", whose components transform by formula (4), as soon as the coordinate systems transform with the help of motion. In other words, these vectors, when depicted as arrows do not change when we transform the coordinate systems with the help of motion; when transforming the coordinate systems with the help of reflection components of such a vector do not transform anymore by formula (4). The vector itself, therefore, transforms in its opposite. Such kind of vectors (which are not physical, by our definition) are called *axial.*

As an example of an axial vector could be given, besides the vector product of two physical vectors, a special vector, called swirl of vector $\mathfrak{a}$ and defined by the equations:

$$\mathfrak{R} = \operatorname{curl} \mathfrak{a};$$

$$\mathfrak{R}_1 = \frac{\partial \mathfrak{a}_3}{\partial x_2} - \frac{\partial \mathfrak{a}_2}{\partial x_2}, \quad \mathfrak{R}_2 = \frac{\partial \mathfrak{a}_1}{\partial x_3} - \frac{\partial \mathfrak{a}_3}{\partial x_1}, \quad \mathfrak{R}_3 = \frac{\partial \mathfrak{a}_2}{\partial x_1} - \frac{\partial \mathfrak{a}_1}{\partial x_2},$$

32

where $\mathfrak{a}$ is a physical vector.

These "vectors" will be discussed in the following chapter.

Later we will see, that axial vectors are similar to the notion anti-symmetric tensor, which will be introduced later on. The most typical axial vectors are the vector product of two polar (physical) vectors.

2. Three-dimensional vectors play a huge role in most of the mechanical and physical situations, concerning the three-dimensional space. When we take into account time, then it is natural to consider phenomena not in the three-dimensional space, but in a fourth-dimensional world (see Introduction), using for description of these phenomena not only three coordinates, but also a fourth coordinate – time; as a result it is necessary to create an apparatus for describing a fourth-dimensional space, analogous to the vector calculus for the case of three-dimensional space. The transition from three dimensions to four dimensions is not more complicated then the general transition to a space of n dimensions. For this reason we will define here the notion of a vector for a space of n dimensions.

The manifold of n dimensions *is a set of n variables $x_1 x_2, \ldots x_n$, which can take on any values in the intervals $(-\infty, +\infty)$.* All these n variables are independent variables. The manifold of n dimensions will be denoted by the symbol M_n.

Variables $x_1, x_2 \ldots x_n$, which form the manifold of n dimensions, are called *coordinates*, the manifold of n given values for the coordinates of the manifold are called a *point* or *element* of the manifold.

If coordinates of one manifold of n dimensions M_n are connected with a given functional dependance with the coordinates of another manifold of n dimensions $\overline{M}_n$, then we say that one manifold transforms in another through a *given point transformation or a given co-ordinate transformation*.

Such transformation of coordinates should posses some properties, limiting its generality; it should, first of all, transform a given point of a manifold in one and only one point of another manifold and vice-verse, whereas this property should hold for all points of this manifold or at least for points of some domain of the manifold, i.e., such values of the coordinates which are belong to given intervals.

The simplest transformation of the coordinates of a manifold is a transformation, in which the new coordinates are connected with the old ones by linear relations. Generalizing formula (1), we will write such a transformation in following way:

$$\overline{x}_i = \alpha_i + \sum_{l=1}^{n} \alpha_l^i x_l, \quad (i = 1, 2, \ldots n), \tag{5}$$

where x_i are coordinates of the first manifold M_n and $\overline{x}_i$ – coordinates of the second (transformed) manifold $\overline{M}_n$.

The requirement, that every point of the second manifold, while transformed, corresponds to one and only one point of the first manifold is expressed obviously by a non-zero determinant:

$$\|\alpha_l^i\| = \begin{vmatrix} \alpha_1^1 & \alpha_2^1 & \dots & \alpha_n^1 \\ \alpha_1^2 & \alpha_2^2 & \dots & \alpha_n^2 \\ \dots & \dots & \dots & \dots \\ \alpha_1^n & \alpha_2^n & \dots & \alpha_n^n \end{vmatrix}.$$

The transformation of the coordinates, expressed by the formulas (5), will be called *linear or affine transformation of the coordinates*.

Obviously, when $n = 3$ formula (5) transforms in formula (1), only with the difference, that the quantities a_l^i will be arbitrary. In this case, when the quantities a_l^i satisfy the equations:

$$\sum_{j=1}^{n} \alpha_r^i \alpha_s^i = \delta_r^s, \quad (r, s = 1, 2, \dots n), \tag{6}$$

where δ_r^s is $+1$ for $r = s$ and 0 for $r \neq s$, we will call the affine transformation (5) *orthogonal affine transformation*; it is obvious that formulas (6) is a direct generalization of equations (2). From formula (5) we can obtain the following equations for orthogonal affine transformations:

$$x_i = \sum_{l=1}^{n} \alpha_i^l (\overline{x}_l - \alpha_l), \tag{7}$$

where $\|\alpha_l^i\|^2 = 1$.

The quantities α_l^i are called *coefficients* of the orthogonal affine, or simply affine, transformations.

Using the notion affine and mainly orthogonal affine transformation of the coordinates, it is not difficult to generalize the definition of a three-dimensional physical vector to a vector of a manifold of n dimension.

Let us assume, that in one of the manifolds of n dimensions, we have a set of n quantities $a_1, a_2, \dots a_n$; these quantities, when a manifold of n dimensions is transformed into another, will be transformed into quantities $\overline{a}_1, \overline{a}_2, \dots \overline{a}_n$; the quantities $\overline{a}_i$ can be chosen arbitrarily, but only for a given choice of the quantities $\overline{a}_1$; the set $a_1, a_2, \dots a_n$ will define a vector in the given manifold.

If for each manifold M_n of n dimensions, transformed from one to another with the help of an orthogonal affine transformation, we *have a set of n quantities $a_1, a_2, \ldots a_n$, transformed into quantities $\overline{a}_1, \overline{a}_2, \ldots \overline{a}_n$ into another manifold $\overline{M}_n$ by the formula:*

$$\overline{a}_i = \sum_{l=1}^{n} \alpha_l^i \, a_l, \quad (i = 1, 2, \ldots n),$$

where α_l^i are coefficients of the affine orthogonal transformation of the manifold M_n into the manifold $\overline{M}_n$, then the set of n quantities a_i defines for each of the given manifold of n dimensions the notion of **affine vector**.

Sometimes, for clarity, the term of affine vector of n-dimensions is used; i.e. the following terms are used: three-dimensional affine vector (which we earlier called simply a vector), four-dimensional affine vector (Vierervector) and so on. We will denote the affine vector for the manifold M_n by bold Latin capital and small letters with a subscript ($\mathbf{A}_i, \mathbf{a}_i$ and so on). We will denote with a line over a bold letter ($\overline{\mathbf{a}}_i$) an affine vector for the manifold $\overline{M}_n$ (obtained from M_n by an orthogonal affine transformation), which corresponds to an affine vector $\mathbf{a}_i$ of the manifold M_n. The set of values $a_1, a_2, \ldots a_n$, which form an affine vector, are called *components of the affine vector*, respectively along the coordinates $x_1, x_2 \ldots x_n$. We will denote them by the same letter as the affine vector, but with a normal, not bold font.

We will give some examples of affine vectors.

Will be call the vector-radius in the manifold M_n of the element $P(x_1, x_2, \ldots x_n)$ relative to the element $P_0(x_1^{(0)}, x_2^{(0)}, \ldots x_n^{(0)})$ the set of quantities $r_1, r_2, \ldots r_n$, which are defined by the equations: $r_i = x_i - x_i^{(0)}$, $(i = 1, 2, \ldots n)$.

It is not difficult to see, that the vector-radius is the affine vector $\mathbf{r}_i$.

We will call the velocity of the element $P(x_1, x_2, \ldots x_n)$ in the manifold M_n, with respect to the variable parameter t, the set of quantities $(v_1, v_2, \ldots v_n)$ defined by the equations; $v_i = \frac{dx_i}{dt}$, where we assume, that x_i are given functions of the parameter t. We will prove that the set of quantities $(v_1, v_2, \ldots v_n)$ is an affine vector v_i:

$$\overline{v}_i = \frac{d\overline{x}_1}{dt}, \quad \overline{x}_i = \alpha_i + \sum_{l=1}^{n} \alpha^i x_l, \quad \frac{d\overline{x}_i}{dt} = \sum_{l=1}^{n} \alpha_l^i \frac{dx_i}{dt}$$

from where

$$\overline{v}_i = \sum_{l=1}^{n} \alpha_l^i \, v_l$$

which proves our assertion.

A set of quantities $f_1, f_2, \ldots f_n$ defined by the equations: $f_i = \frac{\partial f}{\partial x_i}, (i = 1, 2, , \ldots n)$ is called a gradient of a function of the coordinate $f(x_1, x_2, \ldots x_n)$ in the manifold M_n. We will prove that this set of quantities is an affine vector.

We have:

$$\overline{f}_i = \frac{\partial \overline{f}}{\partial x_i} = \sum_{l=1}^{n} \frac{\partial f}{\partial x_l} \cdot \frac{\partial x_l}{\partial \overline{x}_i} = \sum_{l=1}^{n} \alpha_l^i \frac{\partial f}{\partial x_l},$$

because

$$\frac{\partial x_l}{\partial \overline{x}_i} = \alpha_l^i;$$

from where

$$\overline{f}_i = \sum_{l=1}^{n} \alpha_l^i f_l,$$

which proves our assertion.

We should note, that in the last example where we proved the affine vector nature of the gradient, an important role was played by the property of orthogonality of the affine transformation of coordinates, because only by using the conditions of orthogonality, we have:

$$\frac{\partial x_l}{\partial \overline{x}_i} = \alpha_l^i = \frac{\partial \overline{x}_i}{\partial x_l}. \tag{8}$$

Properly generalized and adjusted, this fact will lead us later on to the necessity to divide the vector into two classes: *cogradient* and *contragradient* vectors.

3. Having generalized the notion of a vector for the case of a manifold of n dimensions, we used mainly affine and orthogonal transformation of the coordinates. The necessity, pointed out in the Introduction, to use, when determining the laws of nature, the invariance of these laws under any coordinate transformations (not only affine or orthogonal affine), is related to the complete indifference of nature to systems of coordinates (which could be curvilinear), with the help of which we describe the phenomena of nature – forces us to generalize the notion of a vector for any transformations of coordinates in the manifold M_n.

If the coordinates of the manifold M_n are transformed into the coordinates of manifold $\overline{M}_n$ then this means that between x_i and $\overline{x}_i$ $(i = 1, 2, \ldots n)$ we will have the following relations:

$$\overline{x}_i = \omega_i(x_1, x_2 \ldots x_n), \quad (i = 1, 2, \ldots n) \tag{9}$$

As we already clarified above, while discussing affine transformation, it is necessary that the relations (9) transform a given point of

36

M_n in one and only one point of $\overline{M}_n$ and vice-versa; a similar property should hold at least for a given area of the manifold M_n. Also ω_i should possess some analytical properties, for example to be continuous and have a number of continuous derivatives, at least in the given area of the manifold M_n. Without discussing here the analytical properties of the function ω_i, we will always assume from now on that these functions are always continuous, have as many continuous derivatives as needed in the question that is studied.

It is well known that the condition for a given point M_n to transform in one and only one point $\overline{M}_n$ and vice-versa, is expressed by the non-zero determinant :

$$\left\| \frac{\partial \omega_i}{\partial x_s} \right\| = \frac{D(\omega_1, \omega_2, \dots \omega_n)}{D(x_1, x_2, \dots x_n)},$$

which is called the *Jacobian of the functions* ω_i with respect to the variables x_i. We will always assume that this determinant is different from zero.

Noting that

$$\alpha_s^i = \frac{\partial \overline{x}_i}{\partial x_s} = \frac{\partial x_s}{\partial \overline{x}_i}$$

for orthogonal affine transformations, we can generalize the notion of a vector for any transformation of the coordinates or by the formula:

$$\overline{a}_i = \sum_{s=1}^{n} \frac{\partial \overline{x}_i}{\partial x_s} a_s \quad (i = 1, 2, \dots n)$$

or by the formula:

$$\overline{a}_i = \sum_{s=1}^{n} \frac{\partial x_s}{\partial x_i} a_s, \quad (i = 1, 2, \dots n).$$

The second formula, in the case of a general transformation of the coordinates, will be different from the first, because for the general transformation of the coordinates $\frac{\partial \overline{x}_i}{\partial x_s}$ is not $= \frac{\partial x_s}{\partial \overline{x}_i}$. These two methods of generalization of the notions of affine vector will lead us to two different notions *cogradient* and *contragradient vector*. To distinguish these vectors from each other we will denote a cogradient vector with a superscript, whereas a contragradient vector will be demoted by a subscript.

Based on the facts above and changing some indexes, we formulate the definition of vectors in M_n in the following way.

If for every manifold M_n transforming into one another with the help of point transformations, we have a set of n quantities $T^1, T^2, \ldots T^n$, transforming into the quantities $\overline{T}^1, \overline{T}^2, \ldots \overline{(T^n}$ in another manifolds $\overline{M}_n$ by the formulas:

$$\overline{T}^i = \sum_{s=1}^{n} \frac{\partial \overline{x}_i}{\partial x_s} T^s, \quad (i = 1, 2, \ldots n), \tag{10}$$

then the set of such n quantities T^i defines for each manifold of n dimensions the notion of cogradient vector $\mathbf{T}^i$.

If for every manifold M_n transforming into one another with the help of point transformations, we have a set of n quantities $T_1, T_2, \ldots T_n$, transforming into quantities $\overline{T}_1, \overline{T}_2, \ldots \overline{(T}_n$ in another manifold $\overline{M}_n$ by the formulas:

$$\overline{T}_i = \sum_{s=1}^{n} \frac{\partial x_s}{\partial \overline{x}_i} T_s, \quad (i = 1, 2, \ldots n), \tag{11}$$

then the set of such n quantities T_i defines for each manifold of n dimensions the notion of contragradient vector $\mathbf{T_i}$.

We will denote cogradient and contragradient vectors in the same wat as affine vectors with the only difference that for cogradiant vector we will use superscript and for contragradiant – subscript.

It is evident that an affine vector can be regarded as cogradient as well as contragradient vector (due to the equation $\frac{\partial \overline{x}_i}{\partial x_s} = \frac{\partial x_s}{\partial \overline{x}_i}$).

Later the formulas will be considerably simplified if we introduce the convention, first proposed by Einstein, not to write the sum sign and to imply that in equations with a pair of identical indices we do the summation over these identical indices, where they take on the values from 1 to n. In this convention the formulas (10) and (11) can be rewritten in the following way:

$$\begin{aligned} \overline{T}^i &= T^s \frac{\partial \overline{x}_i}{\partial x_s}, \\ \overline{T}_i &= T_s \frac{\partial x_s}{\partial \overline{x}_i} \end{aligned} \tag{12}$$

When there is no summation over identical indices we will denote that by an asterisk over the formula or it will be specifically mentioned. Based on the above convention we will write the following sum in a much more compact way:

$$\sum_{j=1}^{n} \sum_{k=1}^{n} \sum_{i=1}^{n} A_{ik} B^{ij} C_j^k = A_i k \, B^{ij} C_j^k.$$

38

Formulas (10) and (11) or the equivalent (12) express a vector in $\overline{M}_n$ through a vector in M_n; it is not difficult to obtain from these formulas a reciprocal expression of the vector in M_n through the vector in $\overline{M}_n$, although, of course, this expression will be a simple consequence of the definition of a vector.

As $\overline{x}_i$ are independent variables, then recalling the meaning of the symbol δ_r^s we will have:

$$\frac{\partial \overline{x}_s}{\partial \overline{x}_r} = \delta_r^s,$$

Regarding $\overline{x}_s$, with the help of formula (9), as complex functions of $\overline{x}_r$, depending on $\overline{x}_r$, with the help of $x_1, x_2, \ldots x_n$, we find:

$$\frac{\partial \overline{x}_s}{\partial x_i} \frac{\partial x_i}{\partial \overline{x}_r} = \delta_r^s, \tag{13}$$

exactly in the same way

$$\frac{\partial x_s}{\partial \overline{x}_i} \frac{\partial \overline{x}_i}{\partial x_r} = \delta_r^s. \tag{14}$$

Multiplying the first of the equations (12) by $\frac{\partial x_r}{\partial \overline{x}_i}$ and summing over i from 1 to n and using the formula (14), we find:

$$\overline{T}^i \frac{\partial x_r}{\partial \overline{x}_i} = T^s \frac{\partial \overline{x}_i}{\partial x_s} \frac{\partial x_r}{\partial \overline{x}_i} = T_s \delta_r^s = T^r.$$

In other words, we obtain an expression for a cogradient vector in M_n, through a vector in $\overline{M}_n$. It is clear that this transformation does not contradict the definition of a cogradient vector given above. In a completely analogous way we can deal with contragradient vectors.

It is necessary to mention, that not every affine vector is a vector, in the sense discussed in this section. In fact, the vector-radius of the element P with respect to the element P_0 being an affine vector, will not be a vector anymore, when we introduce general transformations of the coordinates. On the other side, the following easily derived formulas:

$$\overline{v}^i = \frac{d\overline{x}_i}{dt} = \frac{\partial \overline{x}_i}{\partial x_s} \frac{\partial x_s}{\partial t} = v^s \frac{\partial \overline{x}_i}{\partial x_s},$$

$$\overline{f}_i = \frac{\partial \overline{f}}{\partial \overline{x}_i} = \frac{\partial f}{\partial x_s} \frac{\partial x_s}{\partial \overline{x}_i} = f_s \frac{\partial x_s}{\partial \overline{x}_i},$$

show, that the speed of a point in the set of values M_n, relative to the parameter t, is cogradient, whereas the gradient of a function of the point is a contragradiant vector.

It is important to mention, that an infinitely small vector, with components dx_i is a cogradient vector and therefore it is better to write it in the following way: dx^i; but we, however, will keep the old accepted notation, sacrificing the systematic presentation in favour of habit.

4. The introduction of the notions of cogradient and contragradient vectors should not be limited in the rational generalization of vector calculus. The vector product of two vectors already shows, as we have seen above, that this vector product should not be regarded as a vector. It appears natural to think about such a generalization of the notion of vector, which would make it possible to treat the vector product also as some notion that resembles the notion of a vector, but which is a product of a higher rank. When we study the motion of continuous media (hydrodynamics and the theory of elasticity) we introduce elastic stresses, which are notions defined, in the case of our three-dimensional space, by nine quantities whose elements can be written as quantities depending on two indices: i, k $(i, k = 1, 2, 3)$.

The transformation of these quantities, when going from one rectangular coordinate system to another, makes it possible to arrive at the general definition of similar quantities called *tensors of a second rank*.

The generalization of the notion of a tensor of the second rank leads us to the definition of tensors of any kind of rank. We will now define this general notion of a tensor through specific examples as well as in the case of the three-dimensional vector calculus and the theory of motion of continuous media.

If for every manifold M_n, going one into another, with the help of point transformations, we have a set of n^r quantities $T^{i_1 i_2 \ldots i_r}$, or $T_{i_1 i_2 \ldots i_r}$, $(i_1, i_2, \ldots i_r = 1, 2, \ldots n)$ transforming into quantities $\overline{T}^{i_1 i_2 \ldots i_r}$, or in $\overline{T}_{i_1 i_2 \ldots i_r}$ in another manifold $\overline{M}_n$, by the formulas:

$$\overline{T}_1^{i_1 i_2 \ldots i_r} = T^{s_1 s_2 \ldots s_r} \frac{\partial \overline{x}_{i_1}}{\partial x_{s_1}} \frac{\partial \overline{x}_{i_2}}{\partial x_{s_2}} \cdots \frac{\partial \overline{x}_{i_r}}{\partial x_{s_r}}, \tag{15}$$

$$\overline{T}_{i_1 i_2 \ldots i_r} = T_{s_1 s_2 \ldots s_r} \frac{\partial x_{s_1}}{\partial \overline{x}_{i_1}} \frac{\partial x_{s_2}}{\partial \overline{x}_{i_2}} \cdots \frac{\partial x_{s_r}}{\partial \overline{x}_{i_r}}, \tag{16}$$

then the set of n^r quantities $T^{i_1 i_2 \ldots i_r}$ defines for a manifold of the n-th dimension the notion of **cogradient tensor of r-th rank** $T^{i_1 i_2 \ldots i_r}$, *and the set of n^r quantities $T_{i_1 i_2 \ldots i_r}$ defines for each manifold of the n-th dimension the notion of* **contragradient tensor of r-th rank** $T\, T_{i_1 i_2 \ldots i_r}$.

40

Tensors will be denoted by bold Latin letters, having the necessary indices, whereas the components of the tensor will be denoted by normal font.

It is clear from the above definition, that the vector (co- or contragradient) is a tensor of the first rank (co- or contragradient). Applying the previous definitions to tensors of the second rank, which very often appear in concrete physical situations, we will rewrite the formulas (15) and (16) in the following way:

$$\overline{T}^{ik} = T^{\alpha\beta} \frac{\overline{x}_i}{\partial x_\alpha} \frac{\partial \overline{x}_k}{\partial x_\beta}$$

$$\overline{T}_{ik} = T_{\alpha\beta} \frac{\partial x_\alpha}{\partial \overline{x}_i} \frac{\partial x_\beta}{\partial \overline{x}_k}.$$

In order to clarify the notion of a tensor, we will consider the general case of motion of continuous medium, for example viscous liquid or elastic body.

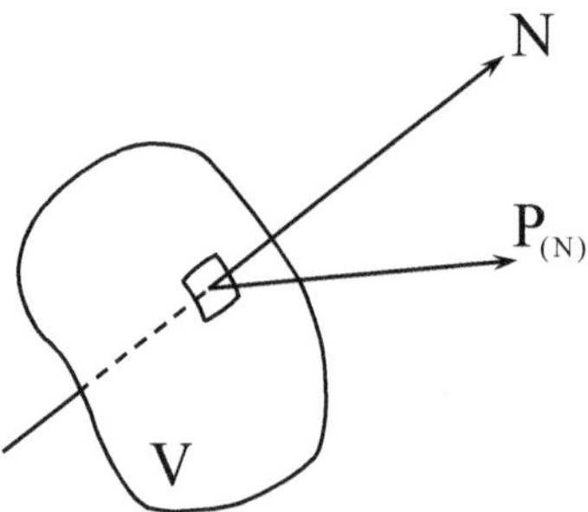

Figure 5

According to d'Alembert's principle, the motion of each system of material points could be derived from the requirement for equilibrium of the acting on each point external forces, forces of inertia and binding forces, to which the system of points is subjected.

Applying this principle to the motion of continuous medium and "mentally cutting out" from that medium (Fig. 5) the volume V, bounded by the surface S, we find that the points of the medium lying outside the volume will exert a force, at each point of the surface S, we find that the points outside the volume will exert at each point of the surface S a force, which is proportional to the area of the element dS of this surface and dependent: 1) on the coordinate of the considered point and 2) on the orientation of the element dS, in other words, on the direction N of its outwards normal. Denoting the vector of this force, taken for a unit area of the surface, by $\mathbf{P}_{(N)}$, we will call it the stress on the area dS with a normal N. Let the components of the stress $\mathbf{P}_{(N)}$ on the axes of the rectangular coordinates be:

$$P_{(N)1},\ P_{(N)2},\ P_{(N)3}.$$

Let us consider the the surface element dS, which are perpendicular to the coordinate axes, we find three stresses $\mathbf{P}_1, \mathbf{P}_2, \mathbf{P}_3$ with the following components on the coordinate axes:

$$\mathbf{P}_1\ :\ P_{11},\ P_{12},\ P_{13};$$

$$\mathbf{P}_2\ :\ P_{21},\ P_{22},\ P_{23};$$

$$\mathbf{P}_3\ :\ P_{31},\ P_{32},\ P_{33}.$$

Considering an infinitesimal tetrahedron whose three sides are formed by the three coordinate planes, and the fourth – by some plane, whose normal N forms angles $\widehat{N, x_1};\ \widehat{N, x_2};\ \widehat{N, x_3}$ with the coordinate axes, and we will obtain the following equation:

$$P_{(N)k} = P_{1k}\,\cos(\widehat{N, x_1}) + P_{2k}\,\cos(\widehat{N, x_2}) + P_{3k}\,\cos(\widehat{N, x_3}),$$

$$(k = 1, 2, 3),$$

$$P_{ik} = P_{ki}, \quad (i, k = 1, 2, 3).$$

These formulas enable us to consider P_{ik}, as components of a tensor, possessing tensor properties, at least for the orthogonal affine transformation of the coordinates. In order to form such kind of tensor, it is necessary to define its components in any of the manifolds M_n, transforming from one into another, with the help of coordinate transformations. We will restrict ourselves only to manifolds M_3, transforming into one another with the help of orthogonal affine transformation of coordinates. We will define the components of our tensor only for such manifolds and will obtain in such a way a notion not of a general but of *affine* tensor of the second rank.

We will assign a rectangular coordinate system to every manifold M_3 with axes x_1, x_2 and x_3. If the manifold $\overline{M}_3$ is obtained from M_3 with the help of an orthogonal affine transformation, then, as the formulas of this transformation show, they are reduced to transformations of systems of rectangular coordinates. It is not difficult to see, that

$$\cos(\widehat{\overline{x}_i, x_l}) = \frac{\partial \overline{x}_i}{\partial x_l} = \frac{\partial x_l}{\partial \overline{x}_i}.$$

We will define the tensor $\mathbf{P}_{jk}$ in M_3 in such a way that its component P_{ik} which is the projection on the axis x_k of the stress on the surface element perpendicular to the axis x_i. Under the given condition in the transformed $\overline{M}_3$, $\overline{P}_{ik}$ will be the projection of the stress $\overline{\mathbf{P}}_i$ on the axis $\overline{x}_k$; we denote the projection of the stress $\mathbf{P}_i$ on the axis x_k (the old axis!) by $\overline{P}_{i,k}$, then applying the above formula for $P_{(N)k}$ and choosing N, directed along the axis $\overline{x}_i$ (the new one!), we obtain the following formula:

$$\overline{P}_{i,k} = P_{1k}\,\cos(\widehat{\overline{x}_i, x_1}) + P_{2k}\,\cos(\widehat{\overline{x}_i, x_2}) + P_{3k}\,\cos(\widehat{\overline{x}_i, x_3}) = P_{\alpha k}\,\cos(\widehat{\overline{x}_i, x_\alpha}).$$

On the other side, it is obvious, that $\overline{P}_{ik}$ – the projection of the stress $\overline{\mathbf{P}}_i$ on the axis $\overline{x}_k$ can be obtained from the projection of this stress on the old coordinate axes by the following formula:

$$\overline{P}_{ik} = \overline{P}_{i\beta}\,\cos(\widehat{x_\beta, \overline{x}_k}).$$

Replacing the index k by β in the previous formula we find:

$$\overline{P}_{ik} = P_{\alpha\beta}\,\cos(\widehat{\overline{x}_i, x_\alpha})\,\cos(\widehat{\overline{x}_k, x_\beta}), \quad (i, k = 1, 2, 3),$$

or, recalling the value of $\cos(\widehat{x_i, x_l})$:

$$\overline{P}_{ik} = P_{\alpha\beta}\,\frac{\partial x_\alpha}{\partial \overline{x}_i}\,\frac{\partial x_\beta}{\partial \overline{x}_k}, \quad (i, k = 1, 2, 3).$$

Changing indices and putting $P_{ik} = P^{ik}$, we find:

$$\overline{P}^{ik} = P^{\alpha\beta}\,\frac{\partial x_\alpha}{\partial \overline{x}_i}\,\frac{\partial x_\beta}{\partial \overline{x}_i}, \quad (i, k = 1, 2, 3,).$$

Hence $\mathbf{P}_{ik} = \mathbf{P}^{ik}$ is affine tensor of the second rank in M_3; as was expected, regardless of whether we will regard this affine vector as co- or contragradient.

As the above tensor is defined only for orthogonal affine transformations in M_3, then one cannot talk about its tensor nature in a general sense. The tensor $\mathbf{P}_{ik} = \mathbf{P}^{ik}$ is called the *tensor of elastic stresses*.

It is not difficult to show that by denoting the components of the velocity $\mathfrak{b}$ on the coordinate axes by $\mathfrak{b}_1$, $\mathfrak{b}_2$, $\mathfrak{b}_3$, we could form a special affine tensor $\mathfrak{p}_{il}$, by using the following relations:

$$p_{ii} = \frac{\partial v_i}{\partial x_i}, \quad p_{ik} = \frac{\partial v_i}{\partial x_k} + \frac{\partial v_k}{\partial x_i}, \quad i \neq k \quad (i, k = 1, 2, 3).$$

This affine tensor $\mathfrak{p}_{ik} = \mathfrak{p}^{ik}$ is called the *tensor of elastic deformations*. The whole theory of elasticity is based on the linear relations between the components of the affine tensor of elastic stress and the components of the affine tensor of elastic deformations.

In the next chapter we will discuss a number of examples not only of affine, but of general tensors. Here we will only note that it is very simple to form a co- or contragradient tensor of any rank by setting in each M_n all components of a tensor to be equal to zero; this pretty trivial tensor will be called *zero tensor*.

5. From the notion of co- or contragradient tensor it is not difficult to pass to the notion of the so called *mixed tensor*, which changes under coordinate transformations like a cogradient tensor, with respect to some indices, and like a contragradient tensor with respect to other indices. Later we will see that this kind of mixed tensors will considerably simplify the calculations.

If for each of the manifolds M_n transforming in one another with the help of transformations (9) we have a set of n^{p+q} quantities

$$T^{i_1\,i_2\ldots i_p}_{j_1\,j_2\ldots j_q} \quad (i_1, i_2 \ldots i_p; j_1, j_2, \ldots j_q = 1, 2, \ldots n),$$

transforming into quantities $\overline{T}^{i_1\,i_2\ldots i_p}_{j_1\,j_2\ldots j_q}$ in another manifold $\overline{M}_n$, respectively by the formulas:

$$\overline{T}^{i_1\,i_2\ldots i_p}_{j_1\,j_2\ldots j_q} = T^{s_1\,s_2\ldots s_p}_{r_1\,r_2\ldots r_q}\, \frac{\partial \overline{x}_{i_1}}{\partial x_{s_1}}\, \frac{\partial \overline{x}_{i_2}}{\partial x_{s_2}} \cdots \frac{\partial \overline{x}_{s_p}}{\partial x_{s_p}} \cdot \frac{\partial x_{r_1}}{\partial \overline{x}_{j_1}}\, \frac{\partial x_{r_2}}{\partial \overline{x}_{j_2}} \cdots \frac{\partial x_{r_q}}{\partial \overline{x}_{j_q}}, \quad (17)$$

then the set of n^{p+q} quantities $T^{i_1\,i_2\ldots i_p}_{j_1\,j_2\ldots j_q}$ defines for each manifold of the n-th dimension the notion of a mixed tensor of $(p+q)$-rank $\mathbf{T}^{i_1\,i_2\ldots i_p}_{j_1\,j_2\ldots j_q}$, **which is cogradient with respect to p upper indices and contragradient with respect to q lower indices.**

It is obvious that the components of a mixed tensor of the second rank will transform by the following formula, directly following from the formula (17):

$$\overline{T}^k_i = T^s_r\, \frac{\partial \overline{x}_k}{\partial x_s}\, \frac{\partial x_r}{\partial \overline{x}_i}. \quad (18)$$

We will show that a system of n^2 quantities δ^k_i, introduced above, is a mixed tensor of the second rank. We have:

$$\delta^k_i = 0, \quad i \neq k, \quad \delta^k_k = 1.$$

Similarly, in a new manifold $\overline{M}_n$ we will have:

$$\overline{\delta}^k_i = 0, \quad i \neq k, \quad \overline{\delta}^k_k = 1.$$

44

Let us show, that the introduced system of quantities satisfies the formulas of transformation (18):

$$\overline{\delta}_i^k = \delta_r^s \frac{\partial \overline{x}_k}{\partial x_s} \frac{\partial x_r}{\partial \overline{x}_i} = \frac{\partial \overline{x}_k}{\partial x_s} \frac{\partial x_s}{\partial \overline{x}_i} = \delta_i^k,$$

by using the property of the introduced transformation (see formula (13)). The last equation, by definition, holds identically and this proves that δ_i^k is a mixed tensor of the second rank. We will call the tensor δ_i^k *unit tensor.*

Notice, that the introduced quantities δ_i^k should be regarded neither as cogradient nor as contragredient tensor; we will assume that we have:

$$\delta_{ik} = \overline{\delta}_{ik} = 0, \quad i \neq k,$$

$$\delta_{ik} = \overline{\delta}_{ik} = 1, \quad i = k.$$

Using the formula (16), if we regard δ_{ik} as a contragradient tensor, we should write the following equation:

$$\overline{\delta}_{ik} = \delta_{\alpha\beta} \frac{\partial x_\alpha}{\partial \overline{x}_i} \frac{\partial x_\beta}{\partial \overline{x}_k},$$

which equation gives obviously the wrong formula:

$$\delta_{ik} = \overline{\delta}_{ik} \frac{\partial x_\alpha}{\partial \overline{x}_i} \frac{\partial x_\alpha}{\partial \overline{x}_k}.$$

6. We showed above that many properties of the world should not depend on the coordinate system, with the help of which we describe the world. Vectors and tensors change in a given way when we change the variables of the manifold, i.e. when we carry out a transformation of the coordinates. For the formulation of the laws, with the help of which we express the properties of our world, we should know how, through vectors and tensors, form such expressions, whose forms do not change under coordinate transformations.

We will call **scalar or invariant** *such a function of the components of vectors and tensors, which will not change under any transformation of the manifold of the kind (9).*

Let us clarify this definition. If we have a given function Φ, dependent on the vectors $\mathbf{a}^i$, $\mathbf{b}_i$ and on the tensors:

$$\mathbf{T}^{i_1 i_2 \ldots i_r}, \quad \mathbf{T}_{i_1 i_2 \ldots i_s}, \quad \mathbf{T}^{i_1 i_2 \ldots i_p}_{j_1 j_2 \ldots j_q}.$$

This dependence can be concisely written in the following way:

$$\Phi = \Phi\left(a^i, \ldots b_k, \ldots T^{i_1 i_2 \ldots i_r}, \ldots T_{i_1 i_2 \ldots i_s} \ldots T^{i_1 i_2 \ldots i_p}_{j_1 j_2 \ldots j_q} \ldots\right).$$

If Φ is scalar, then the following equation should hold:

$$\Phi\left(a^i,\ldots b_k,\ldots T^{i_1\,i_2\ldots i_r},\ldots T_{i_1\,i_2\ldots i_s},\ldots T^{i_1\,i_2\ldots i_p}_{j_1\,j_2\ldots j_q},\ldots\right)$$
$$=\Phi\left(\overline{a}^i,\ldots \overline{b}_k,\ldots \overline{T}^{i_1\,i_2\ldots i_r},\ldots \overline{T}_{i_1\,i_2\ldots i_s},\ldots \overline{T}^{i_1\,i_2\ldots i_p}_{j_1\,j_2\ldots j_q},\ldots\right).$$

If the scalar Φ depends only on vectors – $\overline{\mathbf{a}}^i$ and $\overline{\mathbf{b}}^k$ then the above equation can be written in the following way:

$$\Phi(a^i,\ldots b_k,\ldots) = \Phi(\overline{a}^i,\ldots \overline{b}_k,\ldots).$$

The simplest, but a trivial scalar is any constant since in the transition to new variables the constant quantity does not change at all. It is not difficult to give some examples of less trivial scalars.

The quantity $\Phi = a^i b_i$ is a scalar.
We have:

$$\overline{a}^i = a^\alpha \frac{\partial \overline{x}_i}{\partial x_\alpha}, \quad \overline{b}_i = b_\beta \frac{\partial x_\beta}{\partial \overline{x}_i},$$

$$\overline{a}^i\,\overline{b}_i = a^\alpha \frac{\partial \overline{x}_i}{\partial x_\alpha}\, b_\beta \frac{\partial x_\beta}{\partial \overline{x}_i} = a^\alpha\, b_\beta \frac{\partial \overline{x}_i}{\partial x_\alpha}\frac{\partial x_\beta}{\partial \overline{x}_i},$$

but the sum $\frac{\partial x_i}{\partial x_\alpha}\frac{\partial x_\beta}{\partial \overline{x}_i}$ using formula (14) becomes δ^β_α, so we obtain:

$$\overline{a}^i\,\overline{b}_i = a^\alpha\, b_\beta\, \delta^\beta_\alpha = a^\alpha\, b_\alpha = a^i\, b_i,$$

which proves that Φ is a scalar.

By similar but a bit longer calculations it can be shown that the quantity:

$$\Phi = T^{i_1\,i_2\ldots i_p}_{j_1\,j_2\ldots j_q}\, Q^{j_1\,j_2\ldots j_q}_{i_1\,i_2\ldots i_p} \tag{19}$$

is a scalar. Notice that the quantity Φ, written in such a compact form, is actually a $(p+q)$-fold sum over all lower and upper indices.

It is not difficult to form the following scalar from formula (19):

$$\Phi_1 = T^{i_1\,i_2,\ldots i_p}\, Q_{i_1\,i_2\ldots i_p},\; \Phi_2 = T_{j_1\,j_2\ldots j_q}\, Q^{j_1\,j_2\ldots j_q}.$$

Formula (19) can be used to form many scalars or invariants by using some known vectors and tensors. It seems, however, that formula (19) can provide us with a useful criterion for recognizing whether a given set of quantities has tensor nature or not.

Theorem I. *If a set of quantities b_i $(i = 1, 2, \ldots n)$, defined for each of the manifolds M_n, such that for any cogradient vector $\mathbf{a}^i$ the* *quantity $\Phi = a^i\, b_i$ is a scalar, then this set of quantities b_i is a contragredient vector* $\mathbf{b_i}$.

We go from manifold M_n to manifold $\overline{M}_n$ with the help of a transformation of the type (9); a cogradient vector $\mathbf{a}^i$ transforms into $\overline{\mathbf{a}}^i$ where we have by the definition:

$$\overline{a}^i = a^\alpha\, \frac{\partial \overline{x}_i}{\partial x_\alpha}.$$

The quantities b_i transform into $\overline{b}_i$; using the theorem showing that Φ is a scalar, we obtain the formula relating b_i with $\overline{b}_i$:

$$\Phi = a^i\, b_i = \overline{a}^i\, \overline{b}_i = a^\alpha\, \frac{\partial \overline{x}_i}{\partial x_\alpha}\, \overline{b}_i.$$

Replacing in the last expression the summation variable i by β, and the variable α by i we find:

$$a^i\, b_i = a^i\, \overline{b}_\beta\, \frac{\partial \overline{x}_\beta}{\partial x_i}. \qquad (*)$$

As this equation should hold for each a^i, then the coefficients for any a^i (i has a given value), on the right and on the left should be equal, which gives the following equation:

$$b_i = \overline{b}_\beta\, \frac{\partial \overline{x}_\beta}{\partial x_i}.$$

Replacing i with α in this equation, multiplying both its parts by $\frac{\partial x_\alpha}{\partial \overline{x}_k}$ and summing over α from 1 to n we find:

$$b_\alpha\, \frac{\partial x_\alpha}{\partial \overline{x}_k} = \overline{b}_\beta\, \frac{\partial \overline{x}_\beta}{\partial x_\alpha}\, \frac{\partial x_\alpha}{\partial \overline{x}_k} = \overline{b}_\beta\, \delta_k^\beta = \overline{b}_k.$$

The last equation shows, that $\mathbf{b}_k$ is a contragradient vector which we had to prove.

In the the theorem 1 we assumed, that the cogradient vector $\mathbf{a}^i$ is an arbitrary vector, however such an assumption is unnecessarily broad; we used it only for comparison of the coefficients of a^i in the equation $(*)$; this comparison can be done with more limited assumptions regarding a^i.

The vectors $^1\mathbf{a}^i$, $^2\mathbf{a}^i \ldots {}^r\mathbf{a}^i$ *are called* **linear independent**, *if for their constituents the following relation does not exist*;

$$c_1\, {}^1a^i + c_2\, {}^2a^i + \ldots + c_r\, {}^r a^i = 0, \quad (i = 1, 2, \ldots n),$$

where $c_1, c_2 \ldots c_r$ are numbers which are the same for all i and do not become simultaneously zero.

It is clear, that if we have n linear independent vectors ${}^1\mathbf{a}^i$, ${}^2\mathbf{a}^i, \ldots {}^n\mathbf{a}^i$ then the determinant $\|{}^j a^i\|$ is different from zero.

We notice that the given here definition of linear independent cogradient vectors can be extrapolated to a linear independency of contragredient vectors or in general to tensors of any rank.

Let's assume now that the conditions of theorem 1 hold not for any arbitrary congradient vector $\mathbf{a}^i$, but for some n linear independent congradient vectors. It turns out that even in such a formulation theorem 1 is true. The equation $(*)$ will hold for any n linear independent vectors: ${}^j\mathbf{a}^i$ $(j = 1, 2, \ldots n)$; it can be rewritten in the following way:

$$
{}^j a^i\, b_i = {}^j a^i\, \overline{b}_\beta \frac{\partial \overline{x}_\beta}{\partial x_i} \text{or } {}^j a^i\, (b_i - \overline{b}_\beta \frac{\partial \overline{x}_\beta}{\partial x_i}) = 0, \quad j = 1, 2, \ldots n.
$$

Since the determinant $\|{}^j a^i\|$ is different from zero, then the above system of n linear homogeneous equations with n unknowns:

$$
b_i - \overline{b}_\beta \frac{\partial \overline{x}_\beta}{\partial x_i}, \quad (i = 1, 2, \ldots n).
$$

can have as its solution only zeros e.i., $b_i = \overline{b}_\beta \frac{\partial \overline{x}_\beta}{\partial x_i}$ from where the theorem number 1 will follow. So, theorem 1 can be formulated in the following way:

Theorem 2. *If the set of quantities b_i, $(1, 2, \ldots n)$ defined for each of the manifolds M_n, is such that for any of n linear independent cogradient vectors $\mathbf{a}^i$ the quantity $\Phi = a^i\, b_i$ is a scalar then this set of quantities b_i is a contragradient vector $\mathbf{b}^i$.*

In the same way we can prove the following theorem:

Theorem 3. *If the set of quantities b^i, $(i = 1, 2, \ldots n)$ defined for each of the manifolds M_n is such that for any of n linear independent contragradient vectors $\mathbf{a}_i$ the quantity $\Phi = a_i\, b^i$ is a scalar, then this set of quantities b^i is itself cogradient vector $\mathbf{b}^i$.*

Theorem 4. *If the set of n^{p+q} quantities*

$$
T^{i_1\, i_2 \ldots i_p}_{j_1\, j_2 \ldots j_q}, \quad \left(\begin{matrix} i_1\, i_2 \ldots i_p \\ j_1\, j_2 \ldots j_q \end{matrix} = 1, 2, \ldots n \right),
$$

defined for each of the manifolds M_n is such that for any of n^{p+q} linear independent mixed tensors $\mathbf{Q}^{j_1\, j_2 \ldots j_p}_{i_1\, i_2 \ldots i_q}$ the quantity:

$$
\Phi = T^{i_1\, i_2 \ldots i_p}_{j_1\, j_2 \ldots j_q}\; Q^{j_1\, j_2 \ldots j_q}_{i_1\, i_2 \ldots i_p}
$$

48

is a scalar, then this set of n^{p+q} quantities $\mathbf{T}^{i_1\,i_2\ldots i_p}_{j_1\,j_2\ldots j_q}$ is a mixed tensor of $(p+q)$ rank.

With the help of the above theorem we will often prove the tensor nature of this or that expression. We should notice here that the condition for the linear independence of tensors, with the help of which we prove the tensor nature of our quantities, is an essential condition; to ignore it, as it is sometimes done to simplify the calculation, means to make a serious mistake.

The theorems which we proved could be an extremely simple generalization of the case when as a result of numerous summation, we do not obtain a scalar but some tensor. We will see later on what huge importance this process of summation has for tensor algebra.

Theorem 5. *If $\mathbf{T}^{x_1\,x_2\ldots x_p\;k_1\,k_2\ldots k_\rho}_{\beta_1\,\beta_2\ldots\beta_q\;l_1\,l_2\ldots l_\sigma}$ and $\mathbf{Q}^{\beta_1\,\beta_2\ldots\beta_q\;i_1\,i_2\ldots i_p}_{\alpha_1\,\alpha_2\ldots\alpha_p\;j_1\,j_2\ldots j_q}$ are mixed tensors, then the sum:*

$$R^{i_1\,i_2\ldots i_r\;k_1\,k_2\ldots k_p}_{j_1\,j_2\ldots j_s\;l_1\,l_2\ldots l_q} = T^{\alpha_1\,\alpha_2\ldots\alpha_p\;k_1\,k_2\ldots k_\rho}_{\beta_1\,\beta_2\ldots\beta_q\;l_1\,l_2\ldots l_\sigma}\quad Q^{\beta_1\,\beta_2\ldots\beta_q\;i_1\,i_2\ldots i_r}_{\alpha_1\,\alpha_2\ldots\alpha_q\;j_1\,j_2\ldots j_s}$$

is also a mixed tensor $\mathbf{R}^{i_1\,i_2\ldots i_p\;k_1 k_2\ldots k_\rho}_{j_1\,j_2\ldots j_s\;l_1 l_2\ldots l_\rho}$.

In order not to distract our attention with indices we will prove the theorem for a special case:

$$R^i = T^\alpha\, Q^i_\alpha$$

$$\overline{T}^\alpha = T^\beta\,\frac{\partial \overline{x}_\alpha}{\partial x_\beta},\ \overline{Q}^i_\alpha = Q^\sigma_\gamma\,\frac{\partial \overline{x}_i}{\partial x_\sigma}\,\frac{\partial x_\gamma}{\partial \overline{x}_\alpha},$$

$$\overline{R}^i = \overline{T}^\alpha\,\overline{Q}^i_\alpha = T^\beta\,\frac{\overline{x}_\alpha}{\partial x_\beta}\,Q^\sigma_\gamma\,\frac{\partial \overline{x}_i}{\partial x_\sigma}\,\frac{\partial x_\gamma}{\partial \overline{x}_\alpha}$$

$$= T^\beta\,Q^\sigma_\gamma\,\frac{\partial \overline{x}_i}{\partial x_\sigma}\left(\frac{\overline{x}_\alpha}{\partial x_\beta}\,\frac{\partial x_\gamma}{\partial \overline{x}_\alpha}\right)$$

$$= T^\beta Q^\sigma_\gamma\,\frac{\partial \overline{x}_i}{\partial x_\sigma}\,\delta^\gamma_\beta = T^\beta\,Q^\sigma_\beta\,\frac{\partial \overline{x}_i}{\partial x_\sigma} = R^\sigma\,\frac{\partial \overline{x}_i}{\partial x_\sigma}.$$

and so on.

It is self evident that by reversing theorem 5 we will find easily the criterion for the tensor nature of quantities.

Theorem 6. *If the set of quantities $T^{\alpha_1\,\alpha_2\ldots\alpha_p\;k_1 k_2\ldots k_\rho}_{\beta_1\,\beta_2\ldots\beta_q\;l_1 l_2\ldots l_\sigma}$ (all indices take on values $1, 2, \ldots, n$) is defined for each of the manifolds M_n, such that for any of $n^{p+q+r+s}$ linear independent mixed tensors*

$$\mathbf{Q}^{\beta_1\,\beta_2\ldots\beta_q\;i_1\,i_2\ldots i_r}_{\alpha_1\,\alpha_2\ldots\alpha_p\;j_1\,j_2\ldots j_s}$$

the expression:

$$R^{i_1\,i_2...i_r\,\,k_1\,k_2...k_\rho}_{j_1\,j_2...j_s\,\,l_1\,l_2...l_\sigma} = T^{\alpha_1\,\alpha_2...\alpha_p\,\,k_1\,k_2...k_\rho}_{\beta_1\,\beta_2...\beta_q\,\,l_1\,l_2...l_\sigma}\;\; Q^{\beta_1\,\beta_2...\beta_q\,\,i_1\,i_2...i_r}_{\alpha_1\,\alpha_2...\alpha_p\,\,j_1\,j_2...j_s}$$

is a mixed tensor, then the mentioned set of quantities is also a mixed tensor $\mathbf{T}^{\alpha_1\,\alpha_2...\alpha_p\,\,k_1\,k_1...k_\rho}_{\beta_1\,\beta_2...\beta_q\,\,l_1\,l_1...l_\sigma}$.

This theorem is proved in the same way as the previous; as its formulation is bit awkward we will rewrite it in a more special form. For example, if T_{ik_i} is such that for n linear independent cogradient vectors $\mathbf{a}^i$, the expression

$$b_k = T_{ik}\,a^i$$

is a contragradient vector, then we can assert that $\mathbf{T}_{ik}$ is a contragredient tensor.

In the same way, if for n linear independent contragradient vectors $\mathbf{a}_i$ the expression $b^k = T^{ik}\,a_i$ is a cogradient vector, then $\mathbf{T}^{ik}$ is a cogradient tensor; if for n linear independent cogradient vectors $\mathbf{a}^i$ the expression $b^k = T^k_i\,a^i$ is a cogradient vector then $\mathbf{T}^k_i$ is a mixed tensor.

We will apply the theorems discussed above to prove the tensor nature of some expressions, which are very important for the further presentation of the subject.

First we chose some completely defined contragradient tensor of the second rank $\mathbf{a}_{ik}$ such that $a_{ik} = a_{ki}$ and will call it *fundamental tensor* and will denote it by $\mathbf{g}_{ik}$. Its choice will not be restricted by anything in advance except the property of symmetry in the indices; as a fundamental tensor we can choose any contragradient tensor of the second rank, symmetric in its indices.

The determinant, formed by all n^2 components of the fundamental tensor, will be called *fundamental determinant* and will be denoted by g:

$$g = \|g_{ik}\| = \begin{vmatrix} g_{11} & g_{12} & \cdots & g_{1n} \\ g_{21} & g_{22} & \cdots & g_{2n} \\ \cdots & \cdots & \cdots & \cdots \\ g_{n1} & g_{n2} & \cdots & \cdots g_{nn} \end{vmatrix}.$$

Taking the minor D^{ik} of this determinant corresponding to the i-row and the k-column, in other words, the element g_{ik}, assigning this minor the sign $(-1)^{i+k}$ and dividing it by the fundamental determinant g we obtain a system of n^2 quantities g^{jk}

$$g^{ik} = \frac{(-1)^{i+k}\,D^{ik}}{g}.$$

50

These n^2 quantities for which we will prove that they form a cogradient tensor, have the following property following from the theory of determinants:

$$g^{is}\, g_{sk} = \delta_k^i, \quad g^{si}\, g_{ks} = \delta_k^i. \tag{20}$$

In fact:

$$g^{is}\, g_{sk} = \frac{(-1)^{i+s}\, D^{is}}{g}\, g_{sk} = \frac{(-1)^{i+s}\, D^{is}\, g_{is}}{g},$$

but when $i = k, (-1)^{i+s}\, D^{is}\, g_{is} = g$
and when $i \neq k, (-1)^{i+s}\, D^{is}\, g_{is} = 0$
from where the equation (20) follows.

We will prove that the n^2 quantities g^{ik} form a congradient tensor of the second rank $\mathbf{g}^{ik}$.

Choosing an arbitrary cogradient vector $\mathbf{a}^i$ we can say that, on the basis of theorem 5, $a_k = g_{ik}\, a^i$ will be a contragradient vector $\mathbf{a}_k$.

We will show further that by choosing n linear independent cogradient vectors $\mathbf{a}^i$, with the help of previous formulas, we can form n linear independent contragradient vectors $\mathbf{a}_k$; assume the opposite, that the vectors ${}^j\mathbf{a_k}$ will be linear dependent:

$$c_1\, {}^1a_k + c_2\, {}^2a_k + \ldots c_n\, {}^na_k = 0, \quad (k = 1, 2, \ldots n)$$

Multiplying the equation:

$$a_k = g_{ik}\, a^i$$

by g^{lk} and, using formulas (20), we find

$$g^{lk}\, a_k = g^{lk}\, g_{ik}\, a^i = \delta_i^l\, a^i = a^l,$$

in such a way:

$$ {}^ja^l = g^{lk}\, {}^ja_k.$$

Multiplying the relation expressing the linear dependence of ${}^ja_k$ by g^{lk} and summing each term of this relation over k from $k = 1$ to n, we find:

$$c_1\, {}^1a^l + c_k\, {}^2a^l \ldots + c_n\, {}^na^l = 0, \quad (l = 1, 2, \ldots n),$$

i.e. ${}^ja^l$ will be linear dependent vectors, which contradicts our assumption; so, if $\mathbf{a}^i$ are linear independent vectors then $\mathbf{a}_k = g_{ik}\, a^i$ will also be linear independent contragradient vectors.

We already saw, that:

$$a^l = g^{lk}\, a_k,$$

where $\mathbf{a}^l$ is a cogradient vector for any of the linear independent contragradient vectors $\mathbf{a}_k$. Then, according to theorem 7, we will determine easily that g^{lk} or g^{ik} are components of the cogradient tensor $\mathbf{g}^{ik}$

We will call this cogradient tensor *conjugated fundamental tensor.*

In conclusion we will prove the following property of the fundamental and the conjugated fundamental tensors:

$$g^{ik} g_{ik} = n \tag{21}$$

In fact;

$$g^{ik} g_{ik} = \sum_{i=1}^{n} \sum_{k=1}^{n} g^{ik} g_{ik},$$

but:

$$\sum_{k=1}^{n} g^{ik} g_{ik} = \delta_i^i = 1$$

which proves formula (21).

7. Until now we have been operating with scalars and invariants which do not change their form in any transformations of the variables; in some cases, however, it is more convenient to consider only some scalars which have this property not for any coordinate transformations, but only for affine orthogonal transformations – such scalars will be called *affine scalars.*

As an example of affine scalar we will prove that for any M_n the expression:

$$\Phi = (a_1)^2 + (a_2)^2 + \ldots + (a_n)^2,$$

where a_i are the components of the affine vector $\mathbf{a}_i$, is affine scalar.

According to formula (8) we will have:

$$\overline{a}_i = \sum_{l=1}^{n} \alpha_l^i \, a_l,$$

from were

$$(\overline{a}_i)^2 = \left(\sum_{l=1}^{n} \alpha_l^i \, a_l \right)^2 = \sum_{l=1}^{n} (\alpha_l^i)^2 \, (a_l)^2 - \sum_{l,l'} 2 \, \alpha_l^i \, \alpha_{l'}^i a_l a_{l'},$$

where the second sum is over all different combinations l, l' were l is different from l'. Summing over i we will have:

$$\sum_{i=1}^{n} (\overline{a}_i)^2 = \sum_{l=1}^{n} (a_l)^2 \sum_{i=1}^{n} \alpha_l^i \alpha_l^i + \sum_{ll'} 2 \, a_i \, a_{l'} \sum_{i=1}^{n} \alpha_l^i \alpha_{l'}^i,$$

but since

$$\sum_{i=1}^{n} \alpha_r^i \, \alpha_s^i = \delta_r^s,$$

then:

$$\sum_{i=1}^{n} (\bar{a}_i)^2 = \sum_{l=1}^{n} (a_l)^2 \, \delta_l^l + \sum_{l,l'} 2 \, a_l \, a_{l'} \, \delta_{l'}^l = \sum_{l=1}^{n} (a_l)^2,$$

which proves our assertion.

All previous theorems could be rewritten in theorems only about affine tensors and therefore it is sufficient to write in them the adjective "affine" before the word tensor or vector.

2   ON GROUPS OF TRANSFORMATIONS

1. In the previous chapter we introduced the notions of affine as well as general tensors; each of these categories of tensors was closely connected with a set of transformations transforming one manifold M_n in another $\overline{M}_n$. The development of the ideas of the connection of the notion of a tensor with the set of those transformations, allows us to modify the notion of a tensor and, in addition to affine and general tensors, to arrive at a definition of tensors which relate, in a special way, this set of transformations of one manifold M_n to another $\overline{M}_n$. As a special type of this kind of tensors will be needed later on, when we will establish the foundations of the special principle of relativity, in this chapter we will discuss, on one hand, the question of classification of point transformations transforming M_n into $\overline{M}_n$, and, on the other hand, the connection of tensors with a given set of transformations.

The classification of point transformations, transforming M_n into $\overline{M}_n$ gives us the possibility to define the notion of a *group* of transformations, a notion playing a fundamental role in contemporary mathematics and apparently starting to influence the field of theoretical physics.

We will assume that every point transformation will be denoted symbolically by a letter T with different indices; i.e., the transformation $\overline{x}_1 = \varphi_i(x_1, x_2, \ldots x_n)$ will be denoted by the symbol T and we will write for brevity:

$$\overline{x}_i = T\, x_i, \quad \overline{M}_n = T\, M_n.$$

We will first apply to M_n the transformation T_1, then to the resulting manifold M_n we will apply the transformation T_2 which transforms it into $\overline{M}_n$. The set of these two transformations will transform M_n into $\overline{\overline{M}}_n$ and can be expressed with one transformation T, transforming each element of M_n into the corresponding element of $\overline{\overline{M}}_n$.

What we just said now can be written symbolically in this way:

$$\overline{x}_i = T_1\, x_i, \quad \overline{M}_n = T_1\, M_n,$$

$$\overline{\overline{x}}_i = T_2\, \overline{x}_i, \quad ,\overline{\overline{M}}_n = T_2\, \overline{M}_n,$$

$$\overline{\overline{x}}_i = T\, x_i, \quad \overline{\overline{M}}_n = T M_n.$$

In other words, if the transformation T is expressed by the formulas:

$$\overline{x}_i = \varphi_{1i}\,(x_1, x_2, \ldots x_n)$$

and the transformation T_2 is expressed by the equations:

$$\overline{\overline{x}}_i = \varphi_{2i}\,(\overline{x}_1, \overline{x}_2, \ldots \overline{x}_n),$$

then the transformation T is expressed by the equations:

$$\overline{\overline{x}}_i = \varphi_i\,(x_1, x_2, \ldots x_n),$$

where

$$\varphi_i\,(x_1, x_2, \ldots x_n) =$$
$$\varphi_{2i}\,(\varphi_{11}\,(x_1, x_2, \ldots x_n),\ \varphi_{12}\,(x_1, x_2, \ldots x_n), \ldots \varphi_{1n}\,(x_1, x_2, \ldots x_n)).$$

We will assume that the defined transformation T, which is a result of the application of, first the transformation T_1, and after that the transformation T_2, is called *a product* of the transformations T_1 and T_2 (taken in a given order, first T_1 and then T_2). We will write the product of the transformation T_1 and T_2 as an ordinary product of letters T_1 and T_2:

$$T = T_2\, T_1$$

The previous formula shows that, in general, the transformations $T = T_2\, T_1$ and $T' = T_1\, T_2$ are different from each other; in fact we will have for T:

$$\varphi_i = \varphi_{2i}\, \varphi_{11}\,(x_1, x_2, \ldots x_n),\ \varphi_{12}\,(x_1, x_2, \ldots x_n), \ldots \varphi_{1n}\,(x_1, x_2, \ldots x_{2n}))$$

and for T' we will analogously have:

$$\varphi'_i = \varphi_{1i}\,[\varphi_{21}\,(x_1, x_2, \ldots x_n),\ \varphi_{23}\,(x_1, x_2, \ldots x_n), \ldots \varphi_{2n}\,(x_1, x_2, \ldots x_n)].$$

It is obvious that only in special cases φ'_i coincides with φ_i, so $\varphi'i$ is different from φ_i and the transformation $T' = T_1\, T_2$ is different from

$T = T_2 T_1$. In other words, the product of transformations does not have in the general case the property of commutativeness.

It is not difficult to extrapolate the above notion of a product of two transformations to a product of any number of transformations:

$$T = T_m T_{m-1} \ldots T_2 T_1 = (T_m T_{m_1} \ldots T_2) T_1.$$

Such kind of a product, generally speaking, will depend on the order of the transformations. The product of equal transformations will define the m-th power of transformations:

$$T = T_1^m.$$

It is clear that in this equation the quantity m is a positive integer; with the help of a special definition we can introduce the notion of *zero and negative integer* power of transformation. Let us assume that a transformation which does not change the elements of the manifold M_n, in other words, defined by the equations:

$$\overline{x}_i = \varphi_i (x_1, x_2, \ldots x_n) = x_i,$$

will be called *identical* transformation and will be denoted by the symbols T_0 or 1:
$$M_n = T_0 M_n.$$

We shall call two transformations T_1 and T_2, whose product gives the identical transformation, *reciprocal*:

$$T_2 T_1 = T_0 = 1.$$

If the transformation T_2 is reciprocal to transformation T_1 then it is easy from the formulas for transformation T_1 to obtain the formulas for transformation T_2.

As $T_2 T_1 = T_0$ then the formulas for T_0 will be:

$$\overline{\overline{x}}_i = x_i,$$

i.e $\varphi_i = x_i$ in other words:

$$\varphi_{2i} [\varphi_{11} (x_1, x_2, \ldots x_n), \; \varphi_{12} (x_1, x_2, \ldots x_n) \ldots \varphi_{1n} (x_1, x_2, \ldots x_n)] = x_i,$$

i.e φ_{1i} are obtained from the solution of the equations:

$$x_i = \varphi_{2i} (\overline{x}_1, \overline{x}_2, \ldots \overline{x}_n),$$

for $\overline{x}_i$ exactly as $\varphi_{2i} (\overline{x}_1, \overline{x}_2, \ldots \overline{x}_n)$ are obtained from the solution of the equations $\overline{x}_i = \varphi_{1i} (x_1, x_2, \ldots x_n)$ for x_i,

From the discussion above (with some additional analytical conditions) we can conclude, that if T_2 is reciprocal to T_1, then T_1 is also reciprocal to T_2, because $T_1 T_2$ will give the identical transformation T_0 as the transformation $T_2 T_1$.

Assume that the transformation T_2 is reciprocal to T_1, obtained, therefore, from resolving the formula of the transformation T_1 for x_i, and T_2 is denoted as (-1)-st power of the transformation of T_1:

$$T_2 = T_1^{-1}.$$

From the above discussion it is obvious that if T_2 is reciprocal to T_1 then T_1 is reciprocal to T_2, in other words:

$$T_1 = T_2^{-1} = (T_1^{-1})^{-1}.$$

Let us denote the identical transformation as the zeroth power of any transformation:

$$1 = T_0 = T_1^0$$

then we will have:

$$T_1 T_1^{-1} = T_1^{-1} T_1 = T_1^0 = T_0 = 1$$

After these considerations it is not difficult to define the negative integer power of the transformation T_1 with the help of the following formulas:

$$T_1^{-m} = (T_1^{-1})^m.$$

From this and the previous formulas it is easy to obtain the following rules of operations on transformations and their powers, which are deduced in the same way as the corresponding rules for ordinary powers are deduced in elementary algebra:

$$T_1^m T_1^n = T_1^n T_1^m = T_1^{m+n}$$

$$(T_1^m)^n = T_1^{mn} = T_1^{nm} = (T_1^n)m,$$

if $T_2 T_1 = T$, then $T_1 = T_2^{-1} T$, $T_2 = T T_1^{-1}$, where m and n are any positive, equal to zero or negative integers.

We will give two simplest examples of "actions" on transformations. Assume that the transformations T_1 and T_2 are given by the formulas:

$$T_1 : \bar{x}_i = x_i + a_i, \quad T_2 : \bar{x}_i = x_i + b_i$$

where a_i, b_i, are some constants; then:

$$T = T_2 T_1 = T_1 T_2$$

will be defined by the formula:

$$T: \ \overline{x}_i = x_i + a_i + b_i$$

In the same way $T = T_1^m$ will be defined by the equation:

$$T_1^m: \ \overline{x}_i + m\,a_i,$$

from where:

$$T_1^{-1}: \ \overline{x}_i = x_i - a_i.$$

As a second example, consider the transformation T_1 and T_2:

$$T_1: \ \overline{x}_i = a_i\,x_i, \quad T_2: \ \overline{x}_i = b_i\,x_i,$$

$$T = T_2\,T_1 = T_1\,T_2: \ \overline{x}_i = a_i\,b_i\,x_i,$$

$$T = T_1^m: \ \overline{x}_i = a_i^m\,x_i,$$

where a_i, b_i are constant numbers.

In order to show that the product of transformations is not commutative, we consider the following transformations T_1 and T_2:

$$T_1: \ \overline{x}_i = x_i + a_i,$$

$$T_2: \ \overline{x}_i = x_i^2,$$

then for $T = T_2\,T_1$ we will have the formulas:

$$T = T_2\,T_1: \ \overline{x}_i = (x_i + a_i)^2,$$

Then as for $T' = T_1\,T_2$ we find the equations:

$$T' = T_1\,T_2: \ \overline{x}_i = x_i^2 + a_i.$$

It is clear from here, that T' and T are different from each other.

Usually we consider not ab arbitrary transformation but a *set* of transformations, which can contain a finite number of transformations or a countless set of transformations. In the latter case the set of transformations can be countable or can form a continuum or can possess cardinality greater than the cardinality of the continuum. Of the set of transformations a special role play the so called *groups of transformations*.

A group of transformation is a set of transformations, having the property to contain in itself any product of transformations, belonging to this set.

58

We will denote groups of transformations with letters, placed in boxes, for example $\boxed{T}$.

If T_1 and T_2 are two transformations of some group $\boxed{T}$ of transformations, then this group contains the transformation $T_2 T_1$ and also the transformation $T_1 T_2$.

The just defined notion of a group of transformations does not imply that the group will contain negative powers of the transformations of the group.

We will call a group of transformations *the group, containing pairwise reciprocal transformations,* if each negative power of a transformation, belonging to the group, in its turn also belongs to the group. Further we will mostly consider groups of transformations having pairwise reciprocal transformations and sometimes we will call such groups (for brevity) simply groups of transformations. It is clear that each of these groups contains the identical transformation $T_0 = T_1 T^{-1}$, whereas the general group may not contain the identical transformation.

A necessary and sufficient condition for the group to be a group, containing pairwise reciprocal transformations, is the condition that the group contains all transformations, which are reciprocal to those that are in the group; in other words if for each T_1, belonging to the group we will have that T^{-1} will also belong to the group, then such a group, which is easy to see from the rules of operation on negative powers, will be a group containing pairwise reciprocal transformations.

We will consider as an example the group of transformations of a special kind. We considered above the set of all point transformations:

$$\overline{x}_i = \omega_i \left(x_1, x_2, \ldots x_n \right),$$

where these transformations possess the property of being resolved for x_i for a certain domain of values of these variables, in other words the functions ω_i in the given domain should possess certain analytical properties (more precisely, they should have, in any case, continuous partial derivatives of the first order with respect to the variable x_i in the specified domains) and the Jacobian:

$$\frac{D\left(\omega_1, \omega_2, \ldots \omega_n\right)}{D\left(x_1, x_2, \ldots x_n\right)} = \frac{D\left(\overline{x}_1, \overline{x}_2, \ldots \overline{x}_n\right)}{D\left(x_1, x_2, \ldots x_n\right)}$$

should be different from zero. We will show, that the set of all point transformations forms a group of point transformations and therefore a group containing pairwise reciprocal transformations.

If $T_1 : \ \overline{x}_i = \omega_{1i}(x_1, x_2, \ldots x_n)$ and $T_2 : \ \overline{x}_i = \omega_{2i}(\overline{x}_1; \overline{x}_2, \ldots \overline{x}_n)$ are two point transformations, then their product $T_2 T_1$ defined by the formulas:

$$\overline{\overline{x}}_i = \omega_i(x_1, x_2, \ldots x_n)$$
$$= \omega_{2i}\big(\omega_{11}(x_1, x_2, \ldots x_n),\ \omega_{12}(x_1, x_2, \ldots x_n), \ldots \omega_{1n}(x_1, x_2 \ldots x_n)\big)$$

will be also a point transformation; in fact ω_i will for a specific domain of values of x_i possess the required analytical properties, moreover, by the theorem of multiplication of Jacobians, we will have:

$$\frac{D(\omega_1, \omega_2, \ldots \omega_n)}{D(x_1, x_2, \ldots x_n)} = \frac{D(\overline{\overline{x}}_1, \overline{\overline{x}}_2, \ldots \overline{\overline{x}}_n)}{D(x_1, x_2, \ldots x_n)}$$
$$= \frac{D(\overline{\overline{x}}_1, \overline{\overline{x}}_2, \ldots \overline{\overline{x}}_n)}{D(\overline{x}_1, \overline{x}_2, \ldots \overline{x}_n)} \cdot \frac{D(\overline{x}_1, \overline{x}_2, \ldots \overline{x}_n)}{D(x_1, x_2, \ldots x_n)}.$$

In other words, the Jacobian of the products of transformations will be a product of the Jacobians of these transformations. In such a way, in the domain where these Jacobians-multipliers will be different from zero, and the Jacobian of the products will also be different from zero; from here it will follow that the transformation $\overline{\overline{x}}_i = \omega_i(x_1, x_2, \ldots x_n)$ will be a point transformation. Therefore $T = T_1 T_2$ belongs to the set of point transformations and, hence, this set is a group.

We will show that the group of point transformations is a group containing pairwise reciprocal transformations. In fact, every point transformation $T_1 : \ \overline{x}_i = \omega_{1i}(x_1, x_2, \ldots x_n)$ of the group has a Jacobian different from zero so (due to the above analytical conditions) the group can be resolved for x_i; the transformation $T_1^{-1} : \ x_i = \omega_{2i}(\overline{x}_1, \overline{x}_2, \ldots \overline{x}_n)$ obtained from the resolution of the previous equations for x_i, will be a point transformation, because it will have the necessary analytical properties and will have the Jacobian:

$$\frac{D(x_1, x_2, \ldots x_n)}{D(\overline{x}_1, \overline{x}_2, \ldots \overline{x}_n)} = \frac{1}{\dfrac{D(\overline{x}_1, \overline{x}_2, \ldots \overline{x}_n)}{D(x_1, x_2, \ldots x_n)}}$$

which is different from zero, hence $T_2 = T_1^{-1}$ will belong to the group of point transformations and therefore this group will contain pairwise reciprocal transformations.

The group of point transformations includes all transformations; it is not difficult, however, to pick out from the group a special set of transformations also forming a group. These special groups of transformations can be regarded as subgroups of the point transformations,

60

where, generally speaking, a set of transformations belonging to $\boxed{T}$ and forming a group in its turn is called a subgroup $\boxed{U}$ of the group $\boxed{T}$.

These subgroups of point transformations include the so called m-parametric groups of continuous transformations, which have been thoroughly studied by Lie.[1]

These m-parametric groups can be defined in the following way.

Let us consider the set of transformations, which, in a given way, depends on m real parameters $q_1, q_2 \ldots q_m$:

$$\overline{x}_i = \varphi_i \left(x_1, x_2, \ldots x_n; \; q_1, q_2, \ldots q_m \right).$$

Assume that all m parameters q_j are included in the expression $\overline{x}_i$ through x_i in a *significant way*, i.e., the previous formulas of transformations of arbitrary q_j should not be expressed through less than m number of parameters.

Such kind of sets of transformations will be called *m-parametric set of transformations.*

If any parametric set of transformations possesses the property to be a group, then such a set of transformations will be called m-parametric group of transformations and will be denoted by the symbol $\boxed{G_m}$.

Each transformation of a given group $\boxed{G_m}$ is completely characterized by the parameters $q_1, q_2, \ldots q_m$. We will denote each transformation of this group corresponding to q_j parameters, by the symbols $T\{q_1, q_2, \ldots q_m\}$, or $T\{qj\}$; if $T_1\{q_j^{(1)}\}$ and $T_2\{q^{(2)})_j\}$ are two transformations of our group then their product $T = T_2 T_1$ will be also a transformation of a group, i.e. there will be values q_j of our parameters such that the transformations, mentioned above, give a transformation $T = T_2 T_1$; it is evident that q_j will depend, on $q_1^{(1)}, q_2^{(2)}, \ldots q_m^{(1)}$;, as well as on $q_1^{(3)}, q_2^{(2)}, \ldots q_m^{(2)}$.

Let us clarify the simplest example of the notion m parametric group. Consider the special kind of affine transformations:

$$\overline{x}_i = x + \alpha_i,$$

where the role of parameters q_j is played by n quantities $\alpha_1, \alpha_2, \ldots, \alpha_n$, which can take on any real values. We will show that the set of the above transformations forms a group $\boxed{G_n}$ – in analogy with three-dimensional transformations of coordinates, this group is called *a group*

---

[1] See the fundamental work: Sophus, Lie, *Theorie der transformationsgruppen,* and also the more elementary exposition of the subject in Lie–Sheffers, *Vorlesungen über kontinuirliche Gruppen.*

of parallel transport. Forming a product of the transformations $T_1\{\alpha_i^{(1)}\}$ and $T_2\{\alpha_i^{(2)}\}$ we will have the following transformation:

$$T\{\alpha_i\} : \ \overline{x}_i = x_i + \alpha_i^{(1)} + \alpha_i^{(2)} = x_i + \alpha_i$$

where

$$\alpha_i = \alpha_i^{(1)} + \alpha_i^{(2)},$$

which proves the group character of our set of transformations. It is not difficult to see that the group of parallel transport has identical transformations and contains pairwise reciprocal transformations. From this group it is easy to select a subgroup which does not contain pairwise reciprocal transformations: for this it is sufficient to limit the quantities α_i by the requirement that they are positive or different from zero. In order to have a group with any parameters, which do not contain pairwise reciprocal transformations, it is sufficient to consider n parametric group of such transformations: $\overline{x}_i = x_i + \alpha_i^2$. This group contains the identical transformation, obtained when $\alpha_i = 0$. We can form a group which contains neither pairwise reciprocal transformations nor the identical transformation. Such a group will be, for example, n-parametric group: $\overline{x}_i = x_i + 1 + \alpha_i^2$, with parameters α_i.

The group of parallel transport is a subgroup of a special group, formed by the set of any affine transformations (general, not only orthogonal), i.e. transformations of the following type:[2]

$$\overline{x}_i = \alpha_i + \alpha_l^i x_l,$$

where the determinant $\|\alpha_l^i\| \neq 0$.

A simple calculation shows, that the set of these transformations forms a group containing a pairwise reciprocal transformations. This group is called affine group and have $n + n^2$ parameters (n parameters α_i and n^2 parameters α_l^i), i.e. the affine group will be denoted by the symbol $\boxed{G_{n+n^2}}$; when $n = 2$ we will have $\boxed{G_6}$, when $n = 3 - \boxed{G_{12}}$, when $n = 4 - \boxed{G_{20}}$.

The affine group of transformations divides first of all into two subgroups; the subgroup of parallel transport in the case when all $\alpha_l^i = 0$, except α_l^i equal to 1, and a second group, called *homogeneous affine group*, in the case when all parameters α_i are equal to zero. The transformations of the homogeneous affine group can be written as:

$$\overline{x}_i = \alpha_l^i \, x_l,$$

---

[2] According to the convention above the sign of summation over l is omitted.

62

These transformations *do not change* the element of the manifold, all of whose coordinates are equal to zero.

The homogeneous affine group has n^2 parameters: $\boxed{G_n^2}$.

The affine group itself is a subgroup of a more general group of transformations, called *collineation* or *projective transformations*. The transformations of collineation are defined by the formulas:

$$\overline{x}_i = \frac{\alpha_i + \alpha_l^i\, x_l}{\alpha_{n+1} + \alpha^{n+1}\, x_l} = \frac{\alpha_i + \alpha_1^i\, x_1 + \alpha_q^i\, x_q + \ldots + \alpha_n^i\, x_n}{\alpha_{n+1} + \alpha_1^{n+1} x_1 + \alpha_2^{n+1}\, x_2 + \ldots + \alpha_n^{n+1}\, x_n}$$

where $\alpha_i, \alpha_2^i (i = 1, 2, \ldots n + 1, l = 1, 2, \ldots n)$ are parameters of the transformations.

A simple calculation shows that the above set of transformations forms a group containing pairwise reciprocal transformations. Counting the parameters α_i, α_l^i we find that their number will be equal to $n + n^2 + n + 1$, but dividing the nominator and denominator in the transformations by α_{n+1} we will find, that not all of these parameters are significant. The significant parameters will contain one parameter less – they will be the following parameters $\frac{\alpha_1}{\alpha_{n+1}}$, $i = 1, 2, \ldots n$, $\frac{\alpha_l^i}{\alpha_{n+1}}$, $(i = 1, 2, \ldots n+1, l = 1, 2, \ldots n)$, in such a way, the collineation will be $(2n + n^2)$-parametric group $\boxed{G_{2n+n^2}}$; for $n = 2$ we will have $\boxed{G_8}$, for $n = 3$ and 4 we will, respectively, obtain $\boxed{G_{15}}$ and $\boxed{G_{24}}$.

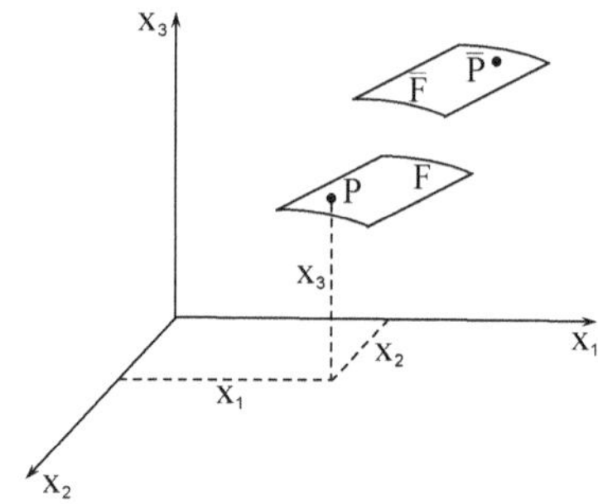

Figure 6

In order to visualize the introduced notions, we consider groups in M_3 where the rectangular coordinates of some point P of our space (see Fig. 6) will be regarded as the numbers x_1, x_2, x_3 of an element of M_3.

Each point transformation of the elements of M_3 can be depicted as relating a point P to one and only one point $\overline{P}$ of space (see Fig. 6). If points P before the transformation have formed a surface F, then after the transformation these points will form another surface $\overline{F}$ and we will say that the surface $\overline{F}$ has been obtained from the surface F with the help of the point transformation or we say that F moved into $\overline{F}$ with the help of point transformations.

The theory of continuous transformations proves, that if we look for the most general group of transformations which transforms planes into other planes, then such group will be *collineation* in M_3, i.e. some $\boxed{G_{15}}$.

It is easy to characterize geometrically the specific properties of the subgroups of collineation, namely – the groups of affine transformations. In general collineation it will be always possible to find such a plane at a finite distance from the origin of coordinates, which will be moved to infinity and will become the so called infinitely distant plane. The affine group will be, as it is clarified in the general theory of group transformations, such collineation, where the *infinitely distant plane remains infinitely distant*, i.e., under transformations this special collineation (affine group) does not change. Of course, we can also consider such subgroups of collineation, in which a certain plane at a finite distance will not change at all under the transformations of this subgroup.

If we studied groups of point transformations not in M_3 but in M_2 then it would be sufficient to go from space to a plane. In this case the collineation is the most general type of finite transformations, transforming any straight line on our plane representing M_2 in some other straight line. The affine group would have been such collineation, which the infinitely distant plane would have remained unchanged. At the end the homogeneous affine group would have been such affine group, whose coordinate origin would have been unchanged.

2. For the clarification of the structure of each group of transformations a significant role play the identification of the subgroups of the given group. These subgroups have often independent importance and could be applied, as we will see further, to a number of questions in geometry and physics. The properties of each group and the identification of its corresponding subgroups significantly depend on the character of two notions, assigned to each group, namely, on the character of the *invariants* of the group and the *invariant equations* of the group. We will confine ourselves to the identification only of the special and the simplest type of invariants and invariant equations of the groups.

We call invariant group $\boxed{T}$ such a function of the coordinates $x_i^{(1)}$ and $x_i^{(2)}$ of any two elements of M_n, which does not change its form

64

under any transformations of the group.

If $I = I(x_1^{(1)}, x_2^{(1)}, \ldots x_n^{(1)}; x_1^{(2)}, x_2^{(2)}, \ldots x_n^{(2)})$ will be the invariant $\boxed{T}$, then under all transformations, belonging to our group, it will not change, that is:

$$I(\overline{x}_1^{(1)}, \overline{x}_2^{(1)}, \ldots \overline{x}_n^{(1)}; \overline{x}_1^{(2)}, \overline{x}_2^{(2)}, \ldots \overline{x}_n^{(2)})$$
$$= I(x_1^{(1)}, x_2^{(1)}, \ldots x_n^{(1)}; x_1^{(2)}, x^{(1)}, x_2^{(2)}, \ldots x_n^{(2)}).$$

It is clear that the set of *all* transformations, under which a given function of the coordinates $I(x_1^{(1)}, x_2^{(1)}, \ldots x_n^{(1)}; x_2^{(1)}, x_2^{(2)}, \ldots \ldots x_n^{(2)})$ does not change its form, forms a group. In fact, if the transformations T_1 and T_2 do not change the form of I, then I will not change its form under the transformations T_2 and T_1 either, i.e., the transformations T_2 and T_1 will belong to the same set as T_1 and T_2, hence this set will be a group. It is easy to see that for the group of parallel transport we can form n invariants of the type;

$$I_i = x_i^{(2)} - x_i^{(1)}.$$

We will call invariant equation of the group $\boxed{T}$ *such an equation, containing the coordinates* $x^{(1)i}$ *and* $x_i^{(2)}$ *of any two elements of* M_n:

$$\Upsilon(x_1^{(1)}, x_2^{(1)}, \ldots x_n^{(1)}; x_1^{(1)}, x^{(2)}, \ldots x_n^{(2)}) = 0,$$

which has these and only these solutions which are also solutions of the equation:

$$\Upsilon(\overline{x}_1^{(1)}, \overline{x}_2^{(1)}, \ldots \overline{x}_n^{(1)}; \overline{x}_1^{(2)}, \overline{x}_2^{(2)}, \ldots \overline{x}_n^{(2)}) = 0,$$

regarded as an equation for $x_i^{(1)}$ *and* $x_i^{(2)}$.

It is clear, that the set of transformations, which leaves a given equation invariant, is a group of transformations.

We will discuss the simplest examples, clarifying the difference between invariant and invariant equations of a given group.

Let us first identify the affine subgroup in M_2 having the invariant $I = (x_1^{(2)} - x_1^{(1)})^2 + (x_2^{(2)} - x_2^{(1)})^2 = 0$ and after that the affine subgroup, having as its invariant equation the following equation:

$$I = (x_1^{(2)} - x_1^{(1)})^2 + (x_2^{(2)} - x_2^{(1)})^2 = 0$$

For the solutions of the first part we will write explicitly the formulas of transformation of the affine group in M_2:

$$\begin{aligned}
\overline{x}_1 &= \alpha_1 + \alpha_1^1\, x_1 + \alpha_2^1\, x_2 \\
\overline{x}_2 &= \alpha_2 + \alpha_1^2\, x_1 + \alpha_2^2\, x_2,
\end{aligned} \qquad (*)$$

The condition that I is invariant can be written as:

$$(\overline{x}_1^{(2)} - \overline{x}_1^{(1)})^2 + (\overline{x}_2^{(2)} - \overline{x}_2^{(1)})^2 = (x_1^{(2)} - x_1^{(1)})^2 + (x_2^{(2)} - x_2^{(1)})^2,$$

where this equation should hold for all $x_i^{(1)}$ and $x_i^{(2)}$. Substituting in this equation formula $(*)$, we obtain the following relation:

$$[(\alpha_2^1)^2 + (\alpha_1^2)^2]\,(x_1^{(2)} - x_1^{(1)})^2 + 2[\alpha_1^1\,\alpha_2^1 + \alpha_1^2\,\alpha_2^2]\,(x_1^{(2)} - x_1^{(1)})\,(x_2^{(2)} - x_2^{(1)}) +$$
$$+ [(\alpha_2^1)^2 + (\alpha_2^2)^2]\,(x_2^{(2)} - x_2^{(1)})^2 = (x_1^{(2)} - x_1^{(1)})^2 + (x_2^{(2)} - x_2^{(1)})^2.$$

Comparing the coefficients of powers and products of the quantities $(x_1^{(2)} - x_1^{(1)})$ and $(x_2^{(2)} - x_2^{(1)})$, we obtain the following relations:

$$(\alpha_1^1)^2 + (\alpha_1^2)^2 = 1,$$

$$\alpha_1^1\,\alpha_2^1 + \alpha_1^2\,\alpha_2^2 = 0,$$

$$(\alpha_2^1)^2 + (\alpha_2^2)^2 = 1.$$

These relations will show immediately that the affine subgroup, which we try to identify, will consist of orthogonal affine transformations and will be called *orthogonal affine group* in M_2. We impose on the four parameters α_i^i three relations, which makes it possible to determine all of these parameters in terms of one of them; we will have two arbitrary parameters α_i and therefore the orthogonal affine group will be a three parametric group $\boxed{G_3}$. It is obvious, that this group will corespond to all kinds of transformations of orthogonal coordinates on the plane, depicting M_2.

Let us identify a second subgroup of affine transformations. The equation $\Upsilon(\overline{x}_1^{(1)}, \overline{x}_2^{(1)};\ \overline{x}_1^{(2)}, \overline{x}_2^{(2)}) = 0$ in our case can be written as:

$$(\overline{x}_1^{(2)} - \overline{x}_1^{(1)})^2 + (\overline{x}_2^{(2)} - \overline{x}_2^{(1)})^2 = 0.$$

Transformed to the variables $x_i^{(1)}$ and $x_i^{(2)}$, this equation will become:

$$[(\alpha_1^1)^2 + (\alpha_1^2)^2]\,(x_1^{(2)} - x_1^{(1)})^2 + 2\,[\alpha_1^1\,\alpha_2^1 + \alpha_1^2\,\alpha_2^2]\,(x_1^{(2)} - x_1^{(1)})\,(x_2^{(2)} - x_2^{(1)}) +$$
$$+ [(\alpha_2^1)^2 + (\alpha_2^2)^2]\,(x_2^{(2)} - x_2^{(1)})^2 = 0$$

This equation should have these and only these solutions as the following equation:

$$(x_1^{(2)} - x_1^{(1)})^2 + (x_2^{(2)} - x_2^{(1)})^2 = 0.$$

Both equations can be viewed as quadratic equations for the quantities

$$\xi = \frac{x_1^{(2)} - x_1^{(1)}}{x_2^{(2)} - x_2^{(1)}}$$

because the roots of one of these equations will coincide with the roots of the other equation, their left-hand sides (quadratic trinomial) will differ only by a constant and positive multiplier λ^2; hence comparing the corresponding coefficients of the trinomial, we will have:

$$\begin{aligned}
(\alpha_1^1)^2 + (\alpha_1^2)^2 &= \lambda^2, \\
\alpha_1^1 \alpha_1^2 + \alpha_1^2 \alpha_2^2 &= 0 \\
(\alpha_2^1)^2 + (\alpha_2^2)^2 &= \lambda^2,
\end{aligned} \tag{**}$$

were λ is an arbitrary quantity.

The three equations relating the parameters α_l^i, in fact, due to the arbitrariness of λ, are reduced to two:

$$\begin{aligned}
(\alpha_1^1)^2 + (\alpha_1^2)^2 &= (\alpha_2^1)^2 + (\alpha_2^2)^2, \\
\alpha_1^1 \alpha_2^1 + \alpha_1^2 \alpha_2^2 &= 0
\end{aligned} \tag{***}$$

which impose given restrictions on these parameters; the third equation from $(**)$ could serve as a definition of λ. The equations $(***)$ show that out of four parameters α_l^i only two are arbitrary; in this way the affine subgroup itself will be four parametric subgroup $\boxed{G_4}$. Each of the transformations of this subgroup could be considered as a product of two transformations:

1) The orthogonal affine transformation T_1:

$$\overline{x}_i = \alpha_i + \beta_l^i x_l$$

and

2) the affine transformation of the kind T_2:

$$\overline{x}_i = \lambda x_i,$$

with positive λ; in fact, the transformation $T = T_2 T_1$ will be:

$$\overline{x}_i = \alpha_i + \lambda \beta_l^i x_l.$$

In other words, will have:

$$\alpha_i = \alpha_i, \quad \alpha_l^i = \lambda\beta_l^i, \qquad\qquad (****)$$

where β_l^i as coefficient of an orthogonal affine transformation should satisfy the relations $(*)$.

We will see whether it is possible, when α_l^i satisfies the relations $(**)$, to choose, with the help of the equation $(****)$, such λ and β_l^i, in order that β_l^i be parameters of the orthogonal affine transformation. Choosing for λ the quantity indicated in the formulas $(**)$ and using the equation $(****)$, we find that β_l^i will satisfy the relations $(*)$, i.e., will be parameters of orthogonal affine transformations.

Let the group $\boxed{G_4}$ be called *semi-orthogonal affine group*. Also the subgroup of the affine group of transformations, in which the following equations are defined:

$$\overline{x}_i = \lambda x_i,$$

when $\lambda > 0$, will be called *group of homothetic transformation*[3] The above discussion shows, that *each transformation of the semi-orthogonal affine group is a product of some transformation of affine orthogonal group and some transformation of the group of homothetic transformations.*

On the other side, the product of any transformation of the orthogonal affine group and any homothetic transformation will be a transformation of the semi-orthogonal affine group. This property can be expressed in the following way: *the semi-orthogonal affine group is a product of the orthogonal affine group and the group of homothetic transformations.*

We will denote the orthogonal affine groups by the symbol $\boxed{O_m}$, the semi-orthogonal groups by $\boxed{\Theta_m}$, where m is the number of the parameters of the groups, the group of homothetic transformations by $\boxed{H_1}$ (it will be obviously one-parametric); then the previous assertions can be symbolically written as:

$$\boxed{\Theta_4} = \boxed{O_3}\,\boxed{H_1}$$

The above discussion can be applied to the general case of orthogonal affine transformations in M_n. Considering affine transformations, having the invariant expression:

$$I = (x_1^{(2)} - x_1^{(1)}) + (x_2^{(2)} - x_2^{(1)})^2 + \ldots + (x_n^{(2)} - x_n^{(1)})^2,$$

---

[3]If x_i are coordinates of continuous medium then it is clear, that homothetic transformations stretch this continuous medium isotropically (the same in all directions).

68

we will arrive at the conclusion that the parameters α_l^i should satisfy the relations:

$$\alpha_r^i \alpha_s^i = \delta_r^s, \quad (r, s = 1, 2 \ldots n),$$

which relations defined the coefficients of orthogonal affine transformations. The number of these relations will be $\frac{n(n+1)}{2}$; in such a way out of n^2 parameters α_l^i only $n^2 - \frac{n(n+1)}{2} = \frac{n(n-1)}{2}$ will be arbitrary parameters, the rest $\frac{n(n+1)}{2}$ are expressed through these $\frac{n(n-1)}{2}$ arbitrary parameters, with the help of the previous relations. Adding to these $\frac{n(n-1)}{2}$ parameters also n more parameters α_i, we find, that orthogonal affine transformations in M_n form a group with $\frac{n(n+1)}{2}$ parameters. This group is called *affine orthogonal group* and can be denoted by the symbol $\boxed{O_{\frac{n(n+1)}{2}}}$. For $n = 3$ we will have, obviously, the group $\boxed{O_6}$, which can be interpreted geometrically as a transformation of one rectangular coordinate system in another.

The orthogonal affine group possesses, as we pointed out in the first chapter, the property that the square of the determinant $\|\alpha_l^i\|$ is equal to 1, so the determinant $\|\alpha_l^i\|$ can be equal to ± 1. We will show that the transformations of the affine group for which the determinant $\|\alpha_l^i\|$ is equal to $+1$, will form a subgroup. This subgroup will be called the *group of motion*. Consider two affine transformations T_1 and T_2:

$$T_1 : \bar{x}_i = {}^1\alpha_i + {}^1\alpha_l^i x_l,$$

$$T_2 : \bar{x}_i = {}^2\alpha_i + {}^2\alpha_l^i x_l$$

Forming the determinant from the coefficients of x_l in the transformation $T_2 T_1$, we find that, by the rule of multiplication of determinants, it will be equal to the product of the determinants $\|{}^1\alpha_l^i\|$ and $\|{}^2\alpha_l^i\|$, corresponding to the transformations T_1 and T_2. It follows from here that if T_1 and T_2 belong to transformations, whose determinants are equal to $+1$, then the determinant corresponding to the transformation $T_2 T_1$ will be also $+1$. In other words, the set of orthogonal affine transformations with a determinant $+1$ will be a group of transformations. On the contrary, the set of affine transformation with a determinant equal to -1, *will not* form a group of transformations.

We will call the special transformation of the type:

$$\bar{x}_1 = x_l, \; \bar{x}_2 = x_2, \ldots \bar{x}_{k-1} = x_{k-1} \; \bar{x}_k = -x_k, \; \bar{x}_{k+1} = x_{k+1}, \ldots \bar{x}_n = x_n,$$

symmetry σ or *reflection* with respect to the coordinate x_k. It is obviously, that the symmetry does not form a group, because $\sigma^2 = T_0 = 1$

gives the identical transformation, but if we add the identical transformation T_0 to any symmetry σ, then the set of two transformations σ and T_0 will form a group, because $\sigma^2 = T_0$, $\sigma T_0 = \sigma$, $T_0^2 = T_0$, $T_0\,\sigma = \sigma$. We will call this group *group symmetry* or reflection and will denote it by $\boxed{\sigma}$. It can be easily proved that *the orthogonal affine group, is a product of the group of motion and the group of symmetry.*

Turning to the representation of M_3 in our space we find (see Fig. 6), that symmetry will be the reflection of a point in one of the coordinate planes, as in a mirror, and the group of motion will correspond to transformations corresponding to the motion of a solid body. Both groups will not change the mutual distance

$$d_{12} = \sqrt{(x_1^{(2)} - x_1^{(1)}) + (x_2^{(2)} - x_2^{(1)}) + (x_3^{(2)} - x_3^{(1)})^2}$$

of any two points.

As in the case of M_2, we can define in the general case the notion of group of *affine semi-orthogonal transformations,* having the following equation invariant:

$$(x_1^{(2)} - x_1^{(1)})^2 + (x_2^{(2)} - x_2^{(1)})^2 + \ldots + (x_n^{(2)} - x_n^{(1)})^2 = 0.$$

The conditions imposed on the parameters α_l^i, can be written after calculations, analogous to the calculations in the case of M_2, in the following way:

$$\alpha_r^I \alpha_s^i = \lambda^2 \delta_r^s,$$

where λ is undefined positive parameter. These equations, show, that in the group of semi-orthogonal affine transformations the number of parameters is greater by 1, compared to the group of affine orthogonal transformations, for that reason this group can be denoted by the following symbol: $\boxed{\Theta_{1+\frac{n(n+1)}{2}}}$, for $n = 3$ this group will be $\boxed{\Theta_7}$.

Introducing the notion of a group $\boxed{N_1}$ of homothetic transformations: $\overline{x}_i = \lambda x_i,\ \lambda > 0$, we will have that *the affine semi-orthogonal group is a product of the affine orthogonal group and the group of homothetic transformations* or symbolically:

$$\boxed{\Theta_{1+\frac{n(n+1)}{2}}} = \boxed{O_{\frac{n(n+1)}{2}}} \cdot \boxed{H_1}.$$

Some features of the physical world (we will talk about them at the end of this chapter), force us, while studying the physical world, to use, not only affine orthogonal transformations, having as an invariant a given special quadratic form of the variables $\xi_i = x_i^{(2)} - x_i^{(1)}$:

$$\xi_1^2 + \xi_2^2 + \ldots + \xi_n^2,$$

70

but also other affine transformations, leaving the quadratic form of the general type invariant:

$$a_{ik}\,\xi_i\,\xi_k,$$

where a_{ik} are constant coefficients, and $\xi_i = x_i^{(2)} - x_i^{(1)}$.

We wil call the affine transformations having as an invariant the expression:

$$I = a_{ik}(x_i^{(2)} - x_i^{(1)})(x_k^{(2)} - x_k^{(1)}),$$

where a_{ik}, are constant real numbers, symmetrical with respect to the indices, affine pseudo-orthogonal transformations corresponding to the quadratic form: $a_{ik}\,\xi_i\,\xi_k$.

The set of the affine pseudo-orthogonal transformations, coresponding to the given quadratic form, constitutes a group called *pseudo-orthogonal affine group, coresponding to the given quadratic form.*

As in the case of orthogonal affine transformations, the parameters of the pseudo-orthogonal affine transformations should satisfy the following relations:

$$a_{ik}\,\alpha_r^i\,\alpha_s^i = a_{rs},$$

In fact:

$$\overline{x}_l^{(2)} - \overline{x}_l^{(1)} = \alpha_l^i(x_l^{(2)} - x_l^{(1)}),$$

for this reason:

$$a_{ik}\,(\overline{x}_i^{(2)} - \overline{x}_i^{(1)})\,(\overline{x}_k^{(2)} - \overline{x}_k^{(1)}) = a_{ik}\,\alpha_r^i\,\alpha_s^i\,(x_r^{(2)} - x^{(1)}1_r)\,(_s^{(2)} - x_s^{(1)}).$$

The expression on the left-hand side should be identically equal to:

$$a_{rs}\,(x_r^{(2)} - x_r^{(1)})\,(x_s^{(2)} - x^{(1)}s).$$

Comparing the coefficients of the different products:

$$(x_r^{(2)} - x_r^{(1)})\,(x_s^{(2)} - x_s^{(1)})$$

we find:

$$a_{ik}\,\alpha_r^i\,\alpha_s^k + a_{ik}\,\alpha_s^i\,\alpha_r^k = a_{rs} + a_{sr}.$$

or:

$$(a_{ik} + a_{ki})\,\alpha_r^i\,\alpha_s^i = a_{rs} + a_{sr},$$

from where, due to the symmetry $a_{ik} = a_{ki}$, $a_{rs} = a_{sr}$ we obtain the following equation

$$a_{ik}\,\alpha_r^i\,\alpha_s^i = a_{rs},$$

which we had to prove.

The above relations (due to the symmetry of a_{rs}), will be $\frac{n(n+1)}{2}$, because, as in the case of orthogonal transformations, the group of affine pseudo-orthogonal transformations will have $n + n^2 - \frac{n(n+1)}{2} = \frac{n(n+1)}{2}$ parameters, i.e. will be $\boxed{G_{\frac{n(n+1)}{2}}}$; for $n = 4$ we will have $\boxed{G_{10}}$.

If the affine transformation has the invariant equation:

$$a_{ik}\left(x_i^{(2)} - x^{(1)i}\right)\left(x_k^{(2)} - x_k^{(1)}\right) = 0,$$

where a_{ik} are constant real numbers, symmetric with respect to the indices, then such a transformation will be called *affine semi-pseudo-orthogonal transformation corresponding to the quadratic form:* $a_{ik}\,\xi_i\,\xi_k$.

The set of affine semi-pseudo-orthogonal transformations, corresponding to the given quadratic form, forms a group called *affine semi-pseudo-orthogonal group*, corresponding to the given quadratic form; we will denote this group by the symbol $\boxed{\mathsf{G}_m}$.

The conditions, which are imposed on the parameter α_l^i of the affine semi-pseudo-orthogonal group, can obviously be expressed by the equation:

$$a_{ik}\,\alpha_r^i\,\alpha_s^k = \lambda^2\,a_{rs}$$

where λ is an arbitrary positive parameter; i.e. the affine semi-pseudo-orthogonal group will have one parameter more than the affine pseudo-orthogonal group and will have: $\boxed{G_{1+\frac{n(n+1)}{2}}}$.

Here we come to the conclusion that *the affine semi-pseudo-orthogonal group, corresponding to the given quadratic form, is a product of the affine pseudo-orthogonal group, corresponding to this quadratic form, and the group of homothetic transformations.*

For the case $n = 4$ the affine semi-pseudo-orthogonal group will be: $\boxed{\mathsf{G}_{11}}$.

It is not difficult to see, that the case of orthogonal transformations is obtained when as the coefficient a_{rs} we take the quantities δ_r^s

$$a_r s = \delta_r^s.$$

In this case the above condition of pseudo-orthogonality can be rewritten in the following way:

$$a_{ik}\,\alpha_r^i\,\alpha^k = \delta^k\,\alpha_r^i\,\alpha_s^k = \alpha_r^i\,\alpha_s^i = \alpha_{rs} = \delta_r^s,$$

i.e. it will change into the ordinary condition of orthogonality.

Later we will have to use a special affine pseudo-orthogonal group, called the general Lorentz group; this is the affine pseudo-orthogonal

72

group $\boxed{G_{10}}$ in M_4 corresponding to the following quadratic form:

$$\xi_1^2 + \xi_2^2 + \xi_3^2 - c^2\xi_4^2,$$

$$a_{11} = a_{22} = a_{33} = 1, \quad a_{44} = -c^2, \quad a_{ik} = 0, \quad i \neq k,$$

where c is some positive constant.

3. Above we introduced the notion of tensors (or in the special case of vectors) of two categories. The tensors of the first category possess the property of transforming its components by the formula (17) under any point transformations, transforming M_n into $\overline{M}_n$; the tensors of the second category – the affine orthogonal tensors – possess this property only for affine orthogonal transformations.

It is completely natural and, as we will see in the second part of our book, extremely useful for studying electrodynamics and the special principle of relativity to change the introduced in the previous chapter notion of tensors and connect this notion with the ideas of groups of transformations. Having set this change of the notion of tensor we turn directly to mixed tensors.

If for each of the manifolds M_n transforming one into another, with the help of the transformations of the group $\boxed{T}$, we have a set of n^{p+q} quantities $T^{i_1 i_2 \ldots i_p}_{j_1 j_2 \ldots j_q}$ $(i_1, i_2, \ldots i_p, j_1, j_2, \ldots j_q = 1, 2, \ldots n)$, transforming into the quantity $\overline{T}^{i_1 i_2 \ldots i_p}_{j_1 j_2 \ldots j_q}$ in another manifold $\overline{M}_n$ according to the formulas:

$$\overline{T}^{i_1 i_2 \ldots i_p}_{j_1 j_2 \ldots j_q} = T^{s_1 s_2 \ldots s_p}_{r_1 r_2 \ldots r_q} \frac{\partial \overline{x}_{i_1}}{\partial x_{s_1}} \frac{\partial \overline{x}_{i_2}}{\partial x_{s_q}} \cdots \frac{\partial \overline{x}_{i_p}}{\partial x_{s_p}} \cdot \frac{\partial x_{r_1}}{\partial \overline{x}_{j_1}} \frac{\partial x_{r_2}}{\partial \overline{x}_{j_2}} \cdots \frac{\partial x_{r_q}}{\partial \overline{x}_{j_q}},$$

then the set of n^{p+q} quantities $T^{i_1 i_2 \ldots i_p}_{j_1 j_2 \ldots j_q}$ defines for each of the mentioned manifolds of the n-th dimension a notion of a mixed tensor of the $(p+q)$ rank corresponding to the group of transformations $\boxed{T}$.

The tensors, corresponding to the group of all point transformations will be called *general tensors*, the tensors corresponding to the group of the affine transformations, are called *affine tensors*, the tensors corresponding to affine orthogonal transformations are called *affine orthogonal tensors*, and finally, the tensors corresponding to the group of motions are named *tensors of motion*.

It is obvious that the more narrow a group is the broader the notion of tensor, corresponding to this group, is. If the group $\boxed{T}$ has a subgroup of transformations $\boxed{U}$, then each tensor, corresponding to $\boxed{T}$, will be in the same time corresponding to the subgroup $\boxed{U}$; the reverse of course does not hold, and not every tensor corresponding to the subgroup $\boxed{\text{U}}$ will be a tensor corresponding the group $\boxed{\text{T}}$. Hence the most special type of tensors will be the tensors called general, on the contrary, the tensors of motion will be the most general of the tensors mentioned above.

In order to visualize the previous considerations better, consider several affine orthogonal vectors in M_3. If $\mathfrak{A}_i$ and $\mathfrak{B}_j$ are affine orthogonal vectors in M_3, then the vector product of these vectors, defined for each of M_3 transforming from one into another with the help of transformations of the affine orthogonal group by three numbers:

$$\mathfrak{G}_i = \mathfrak{U}_2\mathfrak{B}_3 - \mathfrak{U}_3\mathfrak{B}_2, \quad \mathfrak{G}_2 = \mathfrak{U}_1\mathfrak{B}_1 - \mathfrak{U}_1\mathfrak{B}_3, \quad \mathfrak{G}_3 = \mathfrak{U}_1\mathfrak{B}_2 - \mathfrak{U}_2\mathfrak{B}_1,$$

will not be, as we saw in the previous section, affine orthogonal vector.

We will show that the three components of the vector product of two affine orthogonal vectors define a vector of motion; in Chapter 1 we related this vector to a specific type of axial vectors. Moving, with the help of affine orthogonal transformation, from M_3 to $\overline{M}_3$, we will have:

$$\overline{\mathfrak{U}}_i = \alpha_l^i \, \mathfrak{U}_l$$
$$\overline{\mathfrak{B}}_i = \alpha_l^i \, \mathfrak{B}_l$$

from where we find:

$$\overline{\mathfrak{G}}_1 = \overline{\mathfrak{U}}_2\overline{\mathfrak{B}}_3 - \overline{\mathfrak{U}}_3\overline{\mathfrak{B}}_2$$
$$= (\alpha_1^2\mathfrak{U}_1 + \alpha_1^2\mathfrak{U}_2 + \alpha_3^2\mathfrak{U}_3)\,(\alpha_1^3\mathfrak{B}_1 + \alpha_2^3\mathfrak{B}_3 + \alpha_3^3\mathfrak{B}_3)$$
$$- (\alpha_1^3\mathfrak{U}_1 + \alpha_2^3\mathfrak{U}_2 + \alpha_3^3\mathfrak{U}_3)\,(\alpha_1^2\mathfrak{B}_1 + \alpha_2^2\mathfrak{B}_2 + \alpha_3^2\mathfrak{B}_3)$$
$$= (\alpha_2^2\alpha_3^3 - \alpha_3^2\alpha_2^3)\,(\mathfrak{U}_2\mathfrak{B}_2 - \mathfrak{U}_3\mathfrak{B}_2) + (\alpha_3^2\alpha_1^3 - \alpha_1^2\alpha_3^3)\,(\mathfrak{U}_3\mathfrak{B}_1 - \mathfrak{U}_1\mathfrak{B}_3)$$
$$+ (\alpha_1^2\alpha_2^3 - \alpha_2^2\alpha_1^3)\,(\mathfrak{U}_1\mathfrak{B}_2 - \mathfrak{U}_2\mathfrak{B}_1),$$

but for the affine orthogonal transformation we have the following formula:

$$\alpha_2^2\alpha_3^3 - \alpha_3^2\alpha_2^3 = \pm\alpha_1^1, \quad \alpha_3^3\alpha_1^2 - \alpha_1^2\alpha_3^3 = \pm\alpha_2^1, \quad \alpha_1^2\alpha_2^3 - \alpha_2^2\alpha_1^3 = \pm\alpha_3^1,$$

where the sign $+$ corresponds to the case of transformations of motion. Confining ourselves only to these transformations, i.e. keeping the sign $-$ we find:

$$\overline{\mathfrak{G}}_1 = \alpha_1^1\,\mathfrak{G}_1 + \alpha_2^1\,\mathfrak{G}_2 + \alpha_3^1\mathfrak{G}_3,$$

In the same way, we will also show for the other components $\overline{\mathfrak{G}}_2$ and $\overline{\mathfrak{G}}_3$ that the vector product of two affine orthogonal vectors is a vector of the group of motions.

74

Again in the same way we can find that the curl is a vector of the group of motions; in general, each axial vector is a vector of the group of motions.

The vector product of two vectors of the group of motion will be a vector of the group of motion, but the vector product of an axial vector and a polar vector, (in other words, of an affine orthogonal vector) will be, in its turn, an affine orthogonal vector. It is not difficult to notice, that the axial vectors will be only a special type of vectors of the group of motion; in fact, the axial vectors are such vectors of the group of motion, whose components are transformed under the transformation of reflection in a given special way.

For the theory of the special principle of relativity we will need tensors, corresponding to a special pseudo-orthogonal affine group, namely the Lorentz group $\boxed{G_{10}}$ mentioned at the end of the previous chapter.

We will call these tensors *tensors of the Lorentz group*; as they will be needed only for M_4, then sometimes we will call these tensors simply four-dimensional tensors and vectors.

4. In conclusion of this section we will consider the general Lorentz group in M_4, which has extremely important significance in the special principle of relativity and in electrodynamics[4].

First we turn to M_2 and try to find a semi-pseudo-orthogonal and pseudo-orthogonal affine group of transformations, corresponding to the quadratic form:

$$\xi_2^1 - c^2 \, \xi_2^2 = 0.$$

In order to avoid unnecessary indices here, assume, that the element M_2 is defined by a pair of numbers (x, t), in other words, assume that $x_1 = x$, $x_2 = t$; we will rewrite the above quadratic form in the following way:

$$\xi^2 - c^2 \, \tau^2 = 0.$$

In such a way, the coordinate x will correspond to ξ, whereas the coordinate t will correspond to τ. In our case $n = 2$, because of $\frac{n(n+1)}{2} = 3$, hence, for the discussed semi-pseudo-orthogonal group we will have $\boxed{\Theta_4}$ and for pseudo-orthogonal group we will have $\boxed{O}_3$. The general affine transformation in our case will contain $n + n^2 = 6$ parameters; we can now write the general affine transformation in M_2 in

---

[4]See Klein, Ueber die geometrischen Grundlagen der Lorenzgrouppe, Jahresbericht der Deutschen Mathematiker Vereinigung, Bd. 19 (1910); the Russian translation of this article is published in сборнике №7 *Новых Идей в Математике* (под редакцией заслуженного проф. А. В. Васильева) (volume No 7 *New ideas in Mathematics* (edited by the distinguish professor A. V. Vassilev)).

this way:

$$\overline{x} = a + a_1\,x + a_2\,t,$$

$$\overline{t} = b + b_1\,x + b_2\,t.$$

Our further task will be to find the relations between six quantities a, a_1 and so on, bringing the affine group to the semi-pseudo-orthogonal or pseudo-orthogonal group mentioned above.

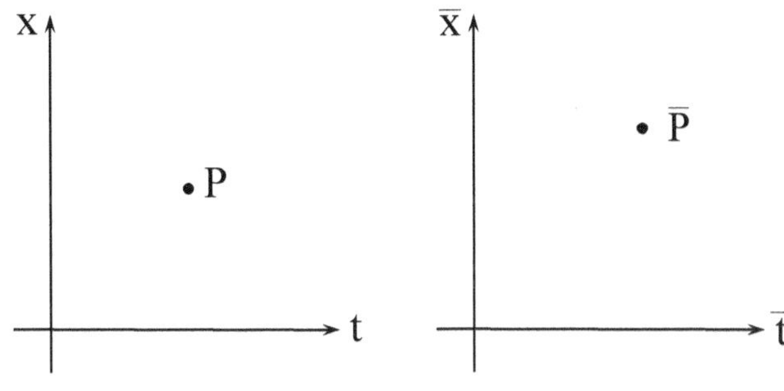

Figure 7

Before solving this elementary algebraic equation, let us see what is the geometrical meaning of this equation. From the geometrical interpretation of the problem it will be easy to go to its physical meaning. We will do it in the chapter devoted to the special principle of relativity.

To each element M_2 we will assign a point P (see Fig. 7) on the auxiliary plane having rectangular coordinates x and t; there, the axis x' is directed vertically up, the axis t is directed horizontally. We will call this plane *image* of M_2; it is self evident that for any $\overline{M}_2$, obtained from M_2 with the help of a point transformation it is possible, as it is shown in the Fig 6, we can form its image. The point transformation makes it possible using the points of one image to construct the points of another one.

Any curve of the image of M_2 will be called a *worldline* or *motion in the image*; in particular, the straight line in the image will be called (see Fig. 8) *uniform motion*; the straight line in the image, which is parallel to the axis t, will be called *rest*, and the straight line of the image, parallel to the axis x will be called an *instant*.

The tangent of the angle of the straight line of the uniform motion with the axis t in the image will be called *velocity* of this motion. Obviously, the velocity of the rest is equal to zero. Concerning the velocity of the instants it is equal to infinity (in the ordinary sense of this word).

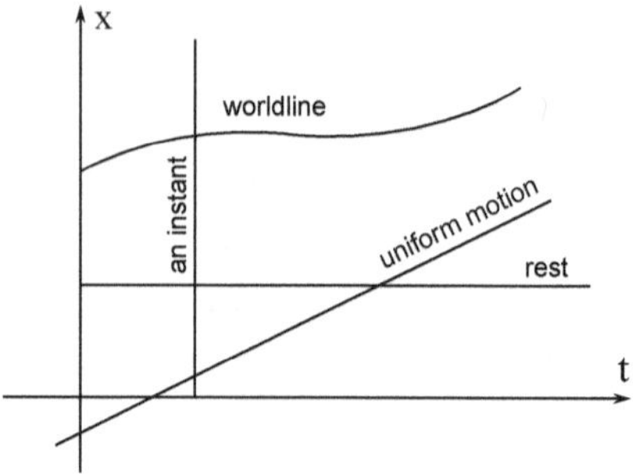

Figure 8

It is clear, that all that terminology is adapted to our usual perceptions for the case when x is an abysses of motion along the straight point and t is time; our figure with its world lines or motions in this case, becomes usual graphical motion (see the graphic of trains). Let us agree that the transition from figure M_2 to the figure $\overline{M}_2$ call *motion of figure*. It is understood, that this term is becoming a simple expression of the straight line motion of solid medium as soon as x and t will be abxice and time, and the point transformation, transforming M_2 into $\overline{M}_2$ will have a special look, which does not change the time;

$$\overline{x} = \omega_1 \left(x_1 t \right)$$

$$\overline{t} = t;$$

in this case the first equation gives as as it is known in Lagrange variable motions of liquid particles, located along on some straight lines and moving down on this straight line[5].

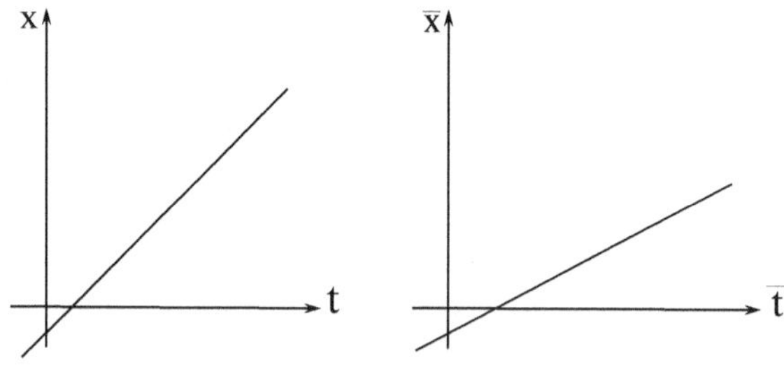

Figure 9

The motion of the figure could be any, we will choose from the variety of figure motions a special type of transfer from one type to

---

[5]Detailed instruction of the interpretation will be cleared up in the chapter, devoted to the special relativity principle

another; *uniformed motion of the figure*[6]. we call such motion of the picture where each uniformed motion on the picture M_2 transfers into uniformed motion on the fig $\overline{M}_2$. From the definition of the uniformed motion of the picture will follow that each straight line on the fig M_2 (see fig. 9) at uniformed motion of the picture, will correspond to the straight line on the picture $\overline{M}_2$ plus, of course, the straight line on finite distance; from this argument it is not difficult, according to the above assertion about the properties affine transformations. We conclude that the most general kind of points transformations where we observe the uniform motion of the figure is affine transformation. I.e. the element $\overline{M}_2$ is connected to the element M_2 when uniform motion corresponds to the figure of affine transformation.

Here we notice (although about it we will talk in detail in the second part in the book), that our uniform motion of the picture reflects the impossibility to discover the ordinary uniform motion of a orthogonal coordinate system by using the law of inertia. It is evident that at the uniform motion of the figure, the rest in the picture $\overline{M}_2$ will go into uniform motion of the picture M_2; the instant on the picture $\overline{M}_2$ will transform into uniform motion in the picture M_2 (see fig. 10).

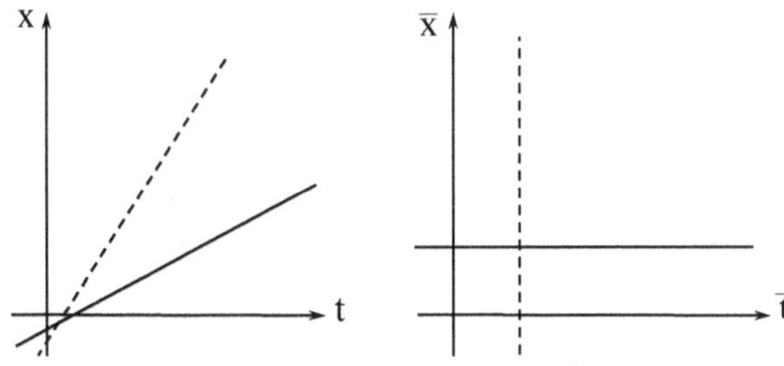

Figure 10

The rest on the picture $\overline{M}_2$, will transfer into uniform motion on the picture M_2 and could change into three kinds of uniform motion; 1) in rest, 2) in instant, and 3) into uniform motion, different from the rest and the instant on the picture $\overline{M}_2$. In the first case we can say, that the uniform motion of the picture is the rest of the picture, the second case we call infinitely fast motion of the picture, the third case will be uniform motion of the picture different from the rest and infinitely fast motion. Analytically, the rest in the picture $\overline{M}_2$ will express with the equation:

$$x = x_0$$

---

[6]The uniform motion of the picture should be different from uniformed motion on the picture.

78

on the picture M_2, the rest according to the formulas of affine transformation, will go in the following straight line (uniform motion):

$$a + a_1 x + a_2 t = \overline{x}_0,$$

the tangent of the angle of the slop of this straight line on the axis t will be equal to $\frac{a_2}{a_1}$ and will not depend on what kind of $\overline{x}_0$ we take, i.e. what kind of rest is considered on the picture $\overline{M}_2$. Let us agree the quantity $\frac{a_2}{a_1}$ call *the speed of the uniform motion of picture* M_2 and denote with v

$$v = -\frac{a_2}{a_1}, \quad a_2 = -a_1 v$$

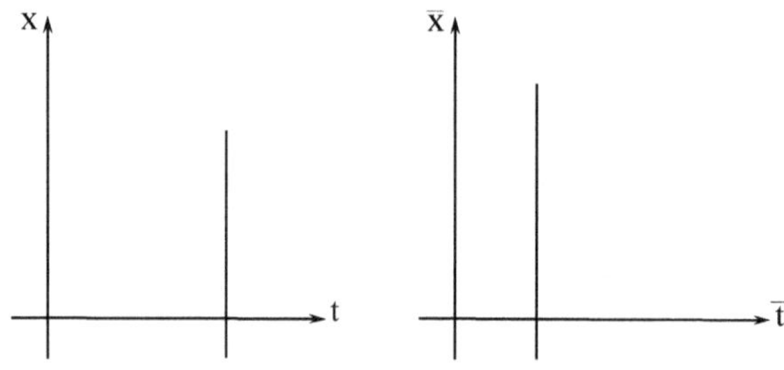

Figure 11

In the case of rest of the picture we will have $v = 0$, in the case of infinitely fast motion of the picture, its speed should not be defined; agree in this case to accept that the speed is equal to ∞, $v = \infty$, expressing trough this symbolical relation the simple equation $a_1 = 0$, corresponding to the case of infinitely fast motion. It is clear, that the introduced by us velocity of the uniform motion of the picture corresponds to the usual notion of velocity in one coordinate system, moving uniformly in a straight line with constant velocity relative to another system.

Generally speaking, the instant of one picture does not transfer to the instant of another picture, but will go in some uniform motion; *we call Galileo's uniform motion* of the picture such uniform motion of the picture, where the instant of one picture goes in the instant of the other picture (see fig. 11). The straight perpendicular axes t on the picture M_2 transfer into straight perpendicular axes $\overline{t}$ on the picture $\overline{M}_2$, in other words in the formulas of transformation we should have; $b_2 = 0$ and these formulas will write in such way:

$$\overline{x} = a + a_1 x + a_2 t,$$
$$\overline{t} = b + b_2 t \tag{22}$$

It is easy to see, that affine transformations (22) form a special group of transformation $\boxed{G}_5$, which we agreed to call *Galilean group*.

From the formulas of transformation is clear, that the uniform motion on the picture M_2 with the velocity v_1 will transfer into uniform motion on the picture $\overline{M}_2$ in general with other velocity v_2 because the straight line of the first uniform motion will be with different slop to the axis t, than this straight line on the picture $\overline{M}_2$ in which it will go.

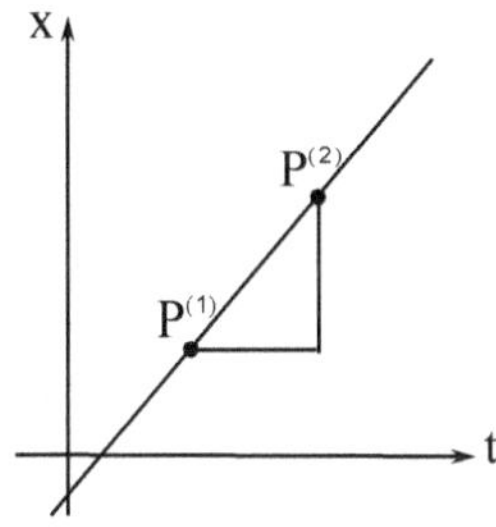

Figure 12

Let us call the uniformed motion on the picture *fundamental* if the numerical value of their velocities does not change at uniformed motion of the picture; in such a way the velocity of the fundamental motion either does not change at all when transferred from fig. M_2 to figure $\overline{M}_2$ or does not change its numerical value but changes its sign. The uniform motion of the picture splits in two classes: the motions of the picture in which exist fundamental uniformed motions can be put in the first class, the rest uniformed motions will be put in the second class. It is clear, that the Galilean uniformed motion of the picture will belong to the first class, because the role of fundamental motion in this case will be played by instances. Let us consider further only the uniformed motions of the pictures, belonging to the first class, i.e. having fundamental motions.

Having the uniform motion on the picture M_2 we can presented it analytically in the following way:

$$x = \nu t = x_o,$$

where the velocity of this motion will be equal to ν.

If on the straight line of the motion (see fig. 12) we consider two points; $P^{(1)}(x^{(1)}, t^{(1)}$ and $P^{(2)}(x^{(2)}, t^{(2)})$, than obviously the velocity of the motion ν will be defined by the formula:

$$\nu = \frac{x^{(2)} - x^{(1)}}{t^{(2)} - t^{(1)}}.$$

80

When the picture M_2 goes to uniformed motion into picture $\overline{M}_2$, than the written above uniformed motion will transfer into this:

$$x = \overline{v}\overline{t} \dots x_o,$$

where

$$\overline{v} = \frac{\overline{x}^{(2)} - \overline{x}^{(1)}}{\overline{t}^{(2)} - \overline{t}^{(1)}},$$

where $(\overline{x}^{(1)}, \overline{t}^{(1)}$ and $(\overline{x}^{(2)}, \overline{t}^{(2)})$ obtained respectively from $(x^{(1)}, t^{(1)}).(x^{(2)}, t^{(2)})$, with the help of affine transformations of the uniform motion of the picture.

When the motion is fundamental, then numerical values ν and $\overline{\nu}$ should be equal. Denoting with the numerical value the velocity of fundamental motion trough c, will have according to defined fundamental motion that if $\pm c$ than $\overline{nu} = \pm c$, where the indices are not connected at all in both equations. Thus the fundamental motion is reduced to the following condition; if we have a relation M_2 on the picture:

$$x^{(2)} - x^{(1)} = \pm c(t^{(2)} - t^{(1)}),$$

then relations on the picture $\overline{M}_2$ should go in the following relations:

$$\overline{x}^{(2)} - x^{(1)})^2 - c^2(t^{(2)} - t^{(1)})^2 = 0.$$

it should be the answer of the equations:

$$(\overline{x}^{(2)} - \overline{x}^{(1)})^2 - c^2(\overline{t}^{(2)} - \overline{x}^{(1)})^2 = 0;$$

from here follows that the unknown affine transformation should have invariant equations of this equation:

$$(x^{(2)} - x^{(1)})^2 - c^2(y^{(2)} - t^{(1)})^2 = 0,$$

i.e. to be *semi-pseudo orthogonal affine transformation, corresponding to quadratic form* $\xi^2 - c^2 t^2 = 0$.

Geometrically, the presence of the fundamental uniform motion with velocity c, obviously requires each set of two straight lines forming with axis t angles with tangents, equal to c and $-c$ (see Fig. 13) in the picture M_2 which have been transformed on the picture $\overline{M}_2$ into set of two straight lines, inclined under the same angles to the axis t. The case of Galilean uniformed motion is, of course, the special case where we have the fundamental motion and which $c = \infty$ both mentioned straight lines are merging into one, perpendicular to the axis t and, which is called an instant.

The discovered by Michelson constancy of the velocity of light, in connection with the long ago established finiteness of this velocity, had given Einstein and Minkowski possibility to consider such an interpretation of the physical phenomena, which led to the special principle of relativity.

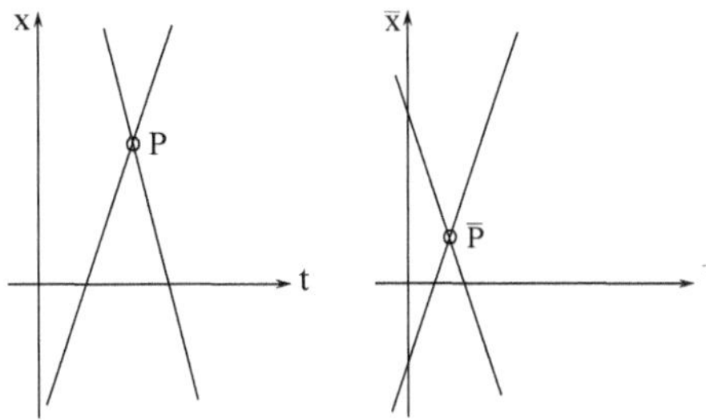

Figure 13

The above discussion had shown the real meaning of the unknown semi-pseudo orthogonal and pseudo-orthogonal affine group, corresponding to the mentioned above quadratic form; Now we will discuss the establishing of these group. At the beginning we defined pseudo-orthogonal affine group; semi-pseudo-orthogonal affine group will be obtained easily by multiplication pseudo-orthogonal by the group of ghemothetical transformations. The value c will be assumed to be finite and different from zero[7].

Using pseudo-orthogonality conditions will have:

$$a_1^2 - c^2 b_1^2 = 1,$$
$$a_1 a_2 - c^2 b_1 b_2 = 0, \qquad (23)$$
$$a_2^2 - c^2 b_2^2 = -c^2$$

Let us clarify first of all, that the assumption $a_1 = 0$ contradicts the mentioned system of equations; in fact, the fist and the second from these equations give:

$$b_1^2 = -\frac{1}{c^2}, \quad b_2 = 0,$$

i.e. it will not be a real number, except that, the definition of the affine transformation, will turn into zero - in other words, duet to these reasons we will get the excluded affine transformation. Hence, a_1 is not

---

[7]The case $c = \infty$ reduces to the subgroup of Galilean groups $\boxed{G}_5$, for which $b_2 = 1$. the case $c = 0$ corresponds the case $a_2 = 0$, i.e. characterizing the rest of picture M_4

$= 0$ and the uniformed motion of our picture will not be infinite fast. Because, a_1 is not $= 0$, than could introduce the parameter $\nu = -\frac{a_2}{a_1}$ and the called by us velocity of the uniformed motion of the picture M_2. Introducing the quantity v, we rewrite the former equations:

$$a_2 = -a_1 v,$$
$$a_1^2 - c^2 b_1^2 = 1,$$

$$a_1^2\, v = c^2\, b_1\, b_2 = 0$$
$$a_1^2\, v^2 - c^2\, b_2^2 = -c^2$$

solving these equations we obtain;:

$$a_1^2 = \frac{1}{1 - \frac{v^2}{c^2}}$$

$$a_2 = -a_1 v_1$$

$$b_1^2 = \frac{\frac{v^2}{c^4}}{1 - \frac{v^2}{c^2}}$$

$$b_2 = -\frac{v}{c^2 b^1 \left(1 - \frac{v^2}{c^2}\right)}$$

denoting $\sqrt{1 - \frac{v^2}{c^2}}$ by β meaning that this quadratic root has a positive value, and denoting ± 1 by ε and ε' we will finally have:

$$a_1 = \frac{\varepsilon}{\beta}, \quad a_2 = -\frac{\varepsilon\, v}{\beta}$$

$$b_1 = \frac{-\varepsilon' \frac{v}{c^2}}{\beta}, \quad b_2 = +\frac{\varepsilon'}{\beta}$$

from where for the pseudo-orthogonal group we find the following formula of transformation:

$$\overline{x} = a + \frac{\varepsilon}{\beta}(x - vt),$$

$$\overline{t} = b + \frac{\varepsilon'}{\beta}\left(t - \frac{v}{c^2}x\right); \tag{24}$$

in this group will enter three parameters a, b and v.

The semi-pseudo-orthogonal group which will satisfy the requirement of existence of a fundamental motion with a finite velocity c, will be obtained from the previous group by multiplying each of the transformations of this group by homeothetical transformation:

$$\overline{x} = \lambda x, \quad \overline{t} = \lambda t;$$

as a result of this multiplication, for obtaining the semi-pseudo-orthogonal group, corresponding to the quadratic form $\xi^2 - c^2\, t^2$, we will have the following transformations:

$$\overline{x} = a + \frac{\varepsilon\lambda}{\beta}(x - vt),$$

$$\overline{t} = b + \frac{\varepsilon\lambda}{\beta}\left(t - \frac{v}{c^2}x\right)$$

from where replacing $\varepsilon\lambda$ with one parameter λ and writing ε instead of ε' we get:

$$\overline{x} = a + \frac{\lambda}{\beta}(x - vt), \overline{t} = b + \frac{\varepsilon\lambda}{\beta}\left(t - \frac{v}{c^2}x\right), \tag{25}$$

where a, b, v, λ are parameters of a group, and $\varepsilon = \pm 1$; hence we have the group $\boxed{\Theta_4}$.

Let us consider the pseudo-orthogonal group in (24) and form a product of two transformations of the group, one with the parameters a', b', v_1, having the values ε_1 and ε_1' for ε and ε', and the other with parameters $a, b", v_2$ having values ε_2 and ε'_2 for ε and ε'; because the considered transformations belong to the group, than their product should be one of the transformations of the group, i.e. should be brought to the kind of (24); let us define the values $a, b, v, \varepsilon, \varepsilon'$ for this product of transformations. So we will have;

$$\overline{x} = a' + \frac{\varepsilon_1}{\beta_1}(x - v_1 t),$$

$$\overline{t} = b' + \frac{\varepsilon_1'}{\beta_1}\left(t - \frac{v_1}{c^2}x\right)$$

$$\overline{\overline{x}} = a" + \frac{\varepsilon_2}{\beta_2}(\overline{x} - v_2\overline{t})$$

$$\overline{\overline{t}} = b" + \frac{\varepsilon_2'}{\beta - 2}\left(\overline{t} - \frac{v_2}{c_2}\overline{x}\right)$$

$$\overline{\overline{x}} = a + \frac{\varepsilon}{\beta}(x - vt)$$

$$\overline{\overline{t}} = b + \frac{\varepsilon'}{\beta}\left(t - \frac{v}{c^2}x\right)$$

84

where $\beta_1 = \sqrt{1 - \frac{v_1^2}{c^2}}, \quad \beta_2 = \sqrt{1 - \frac{v_2^2}{c^2}}, \quad \beta = \sqrt{1 - \frac{v^2}{c^2}}$

Using the four equations, at the left side, we have:

$$\bar{\bar{x}} = a'' + \frac{\varepsilon_2}{\beta_2}(a' - v_2 b') + \frac{\varepsilon_2}{\beta_2}\frac{1}{\beta_1}\left[(\varepsilon_1 + \frac{\varepsilon_1 v_1 v_2}{c^2})x - (\varepsilon_1 v_1 + \varepsilon'_1 v_2)t\right]$$

$$\bar{\bar{t}} = b'' + \frac{\varepsilon'_2}{\beta_2}(b' - \frac{v_2}{c_2}a') + \frac{\varepsilon'_2}{\beta_2}\frac{1}{\beta_1}\left[(\varepsilon'_1 + \frac{\varepsilon_1 v_1 v_2}{c^2})t - \frac{\varepsilon'_1 v_1 + \varepsilon_1 v_2}{c^1}x\right]$$

from these formulas, when compared to the equations at the left in the previous relations easy we obtained:

$$a = a'' + \frac{\varepsilon_2}{\beta_2}(a' - v_2 b'), \quad b = b'' + \frac{\varepsilon'_2}{\beta_2}(b' - \frac{v_2}{c^2}a')$$

$$\frac{\varepsilon}{\beta} = \frac{\varepsilon_2}{\beta_2 \beta_4}(\varepsilon_1 + \frac{\varepsilon'_1 v_1 v_1}{c'}), \quad \frac{\varepsilon v}{\beta} = \frac{\varepsilon_2}{\beta_1 \beta_2}(\varepsilon_1 v_1 + \varepsilon'_1 v_2)$$

$$\frac{\varepsilon'}{\beta} = \frac{\varepsilon'_2}{\beta_1 \beta_2}\varepsilon'_1 + \frac{\varepsilon_1 v_1 v_2}{c_2}) \quad \frac{\varepsilon' v}{\beta c^2} = \frac{\varepsilon'_2}{\beta_1 \beta_2}(\frac{\varepsilon'_1 v_2 + \varepsilon_1 v_2}{c^2})$$

the first two equations give a and b, the rest should define v and quantities ε and ε' having only two values; $+1$ and -1. To solve the previous equations, denote relation $\frac{\varepsilon'_1}{\varepsilon_1}$ with $\eta(\eta = \pm 1)$; divide an equation from the top on the other we get:

$$v = \frac{v_1 + \eta v_2}{1 + \frac{\eta v_1 v_2}{c^2}} \tag{*}$$

dividing an equation from the second line on the other and assume that $\frac{\varepsilon_1}{\varepsilon'_1} = \frac{1}{\eta} = \eta \, (\eta = \pm 1)$ we will obtain the same value for v. We should define only the values ε and in ε' and check whether the above equations are satisfied. These equations could be written in the following way:

$$\frac{\varepsilon}{\beta} = \frac{\varepsilon_1 \varepsilon_2}{\beta_1 \beta_2}\left(1 + \frac{\eta v_1 v_2}{c^2}\right)$$

$$\frac{\varepsilon'}{\beta} = \frac{\varepsilon'_1 \varepsilon'_2}{\beta_1 \beta_2}\left(1 + \frac{\eta v_1 v_2}{c^2}\right)$$

It is not difficult to verify the following identity by direct calculation:

$$\frac{1}{\beta} = \frac{1 + \frac{\eta v_1 v_2}{c^2}}{\beta_1 \beta_2}$$

where v is defined by the formula $(*)$; this identity shows, that the previous equations satisfy if we denote:

$$\varepsilon = \varepsilon_1\varepsilon_2 \quad \varepsilon' = \varepsilon'_1\varepsilon'_2$$

We will show that the pseudo-orthogonal group which we just consider contain pairwise opposite transformations and find the formula, connected these transformations. First of all it is clear that $a = 0, b = 0$, $v = 0$, $\varepsilon = +1$, $\varepsilon' = +1$ we will have identical transformation; let transformations with the parameters, defined above, be reversible; their product will give identical transformation. Without devoting time to the connections a'' and b'' $ca'b'$, we will consider the connections v_2 and v_1. As $v = 0$, we will have;

$$v_1 + \eta v_2 = 0, \quad v_2 = -\eta v_1,$$

On the other side as far as ε and ε' are equal to $+1$ the equations:

$$\varepsilon_1\varepsilon_2 = +1, \quad \varepsilon_1\varepsilon'_2 = +1$$

$$\varepsilon_2 = \varepsilon_1, \quad \varepsilon'_2 = \varepsilon'_1$$

Let us have an additional requirment, that the parameters v_1 and v_2 for opposite transformations would have equal values and are diffident only by the sign, then we will have $\eta = +1^8$ in other words, we find:

$$\varepsilon_1 = \varepsilon'_1$$

and the formulas of transformation will be:

$$\overline{x} = a + \frac{\varepsilon}{\beta}(x - vt)$$

$$\overline{t} = b + \frac{\varepsilon}{\beta}(t - \frac{v}{c^2}x)$$

the mentioned pointed out pseudo- orthogonal group will have subgroup, obtained at $\varepsilon = +1$; this subgroup is named *special group of Lorentz* for M_2; its transformation could be written in the following way;

$$\overline{x} = a + \frac{x - vt}{\sqrt{1 - \frac{v^2}{c^2}}}$$

$$\overline{t} = b + \frac{t - \frac{v}{c^2}x}{\sqrt{1 - \frac{v^2}{c^2}}} \tag{26}$$

---

[8] Here we have the requirement that the velocity of the picture $\overline{M_2}$ would have been the same concerning the quantities and opposite by the sign of the velocity of M_2.

In the chapter of the special relativity principle we will explain why we could avoid the general lorentz group, limiting ourselves only to the special Lorentz group.

Now we are going to discussing the properties of Lorentz group in M_4. *The Lorentz general group M_4* we call pseudo-orthogonal affine group, corresponding to the quadratic formula of the following kind:

$$\xi^2 + \eta^2 + \delta^2 - c^2\tau^2 = 0,\ ^9$$

where c is finite and different from zero constant.

We will call *the expanded Lorentz group M_4* the semi-pseudo-orthogonal affine group, corresponding to the mentioned above quadratic form.

The general Lorentz group in M_4 will have, according to the general theory of pseudo and semi-pseudo-orthogonal groups 10 parameters, the expanded Lorentz group will have 11 parameters. Let us denote the general Lorentz group by the symbol $\boxed{L}_{10}$ and the expanded group of Lorentz by the $\boxed{L'}_{11}$.

It is easy to see, that $\boxed{L'}_{11}$ is a a direct generalization of the corresponding group for M_2. This generalization could be obtained as soon as we spread all the notions of motions, uniform motion, velocity etc. on M_4 which we found for M_2. This generalization would be discussed in detail in the chapter devoted to the special principle of relativity, now we only will point out that the affine group is a result of the demand that every uniform motion in M_4 will be transferred into the uniform motion in $\overline{M}_4$(the same way as it is for the case of the manifolds of two dimensions). The expanded group of Lorentz is the only group of transformation where there is a fundamental uniform motion with the velocity which real value does not change at the transition of M_4 to $\overline{M}_4$

The expanded Lorentz group $\boxed{L'}_{11}$ is the product of $\boxed{L}_{10}$ and the group of homothetiocal transformations; let see at what kind of simpler elements the general Lorentz group could be split. Make the coefficient affine group in M_4 simpler, we will have the following formulas for the transformations of this group $\boxed{G}_{20}$:

---

[9]For simplicity we do not use here the unnecessary indices; the element M_4 will have for coordinates its x, y, z, t which coordinates in the quadratic formula will be the letters $\xi, \gamma, \delta, \tau$.

$$\overline{x} = \alpha = \alpha_1 x + \alpha_2 y + \alpha_3 z + \alpha_4 t.$$
$$\overline{y} = \beta = \beta_1 x = \beta_2 y + \beta_3 z = \beta_4 t.$$
$$\overline{z} = \gamma + \gamma_1 y = \gamma_3 z = \gamma_4 t.$$
$$\overline{t} = \delta + \delta_1 x + \delta_2 y + \delta_3 z + \delta_4 t.$$

the pseudo-orthogonal condition, corresponding to the above written quadratic form, gives us the following relations between a_1, a_2 and so on.

$$(\alpha_1)^2 + (\beta_1)^2 + (\gamma_1)^2 - c^2(\delta_1)^2 = 1.$$
$$(\alpha_2)^2 + (\beta_2)^2 + (\gamma_2)^2 - c^2(\delta - 2)^2 = 1.$$
$$(\alpha_3)^2 + (\beta_3)^2 + (\gamma_3)^2 - c^2(\delta_3)^2 = 1.$$
$$(\alpha - 4)^2 + (\beta_4)^2 + (\gamma_4)^2 - c^2(\delta_4)^2 = -c^2.$$
$$\alpha_2\alpha_3 + \beta_2\beta_3 + \gamma_2\gamma_3 - c^2\delta_2\delta_3 = 0.$$

$$\alpha_3\alpha_1 = \beta_3\beta_1 + \gamma_3\gamma_1 - c^2\delta_3\delta_1 = 0$$
$$\alpha_1\alpha_2 = \beta_1\beta_2 + \gamma_1\gamma_2 - c^2\delta_1\delta_2 = 0$$
$$\alpha_1\alpha_4 = \beta_1\beta_4 + \gamma_1\gamma_4 - c^2\delta_1\delta_4 = 0$$
$$\alpha_2\alpha_4 + \beta_2\beta_4 + \gamma_2\gamma_4 - c^2\delta_2\delta_4 = 0$$
$$\alpha_3\alpha_4 + \beta_3\beta_4 + \gamma_3\gamma_4 - c^2\delta_3\delta_4 = 0$$

From the previous formulas becomes clear, that coordinates x, y, z and the coordinate t of element M_4 we subscribe different role; let us call the coordinates x, y, z *spatial* and the coordinate t *time*.

We show, that *each transformation of Lorentz group* $\boxed{L}_{10}$ *could be formed of the product: 1) someorthogonal affine transformation, which does not change the time coordinate, 2) some Lorentz transformation in M_2, have one spatial and one time coordinate, and 3) some orthogonal affine transformation, also does not changing the time coordinate.*

The above assumption could be formulate the following way. Let some transformations of the group $\boxed{L}_{10}$ transmit M_4 into $\overline{\overline{M}}_4$; we denote the transformation as;

$$\begin{pmatrix} x, & y, & z, & t \\ \overline{\overline{x}}, & \overline{\overline{y}}, & \overline{\overline{z}}, & \overline{\overline{t}} \end{pmatrix}$$

88

Denote the affine orthogonal transformation, transforming x, y, z into $\overline{x}, \overline{y}, \overline{z}$ while not changing t, by the symbols $O\left(\frac{x,\ y,\ z}{\overline{x},\ \overline{y},\ \overline{z}}\right)$; after we denote Lorentz transformation, which transform $\overline{x}, \overline{t}$ into x, t while not changing $\overline{y}$ and $\overline{z}$ by the symbol $\text{Л}\left(\frac{\overline{x},\ \overline{t}}{\overline{\overline{x}},\ \overline{z}}\right)$, and finally, the affine orthogonal transformation, transforming $\overline{\overline{x}}, \overline{\overline{y}}, \overline{\overline{z}}$ into $\overline{\overline{\overline{x}}}, \overline{\overline{\overline{y}}}\ \overline{\overline{\overline{z}}}$ and not changing $\overline{t}$ we denote with the symbol $O'\left(\frac{\overline{\overline{x}},\ \overline{\overline{y}},\ \overline{\overline{z}}}{\overline{\overline{\overline{x}}},\ \overline{\overline{\overline{y}}},\ \overline{\overline{\overline{z}}}}\right)$. This assumption is written in a symbolic way in the following way:

$$L\begin{pmatrix} x, & y, & z, & t \\ \overline{\overline{\overline{x}}}, & \overline{\overline{\overline{y}}}, & \overline{\overline{\overline{z}}}, & \overline{t} \end{pmatrix} = O'\begin{pmatrix} \overline{\overline{x}}, & \overline{\overline{y}}, & \overline{\overline{z}} \\ \overline{\overline{\overline{x}}}, & \overline{\overline{\overline{y}}}, & \overline{\overline{\overline{z}}} \end{pmatrix} \text{Л}\begin{pmatrix} \overline{x}, & \overline{t} \\ \overline{\overline{x}}, & \overline{z} \end{pmatrix} O\begin{pmatrix} x, & y, & z \\ \overline{x}, & \overline{y}, & \overline{z} \end{pmatrix}$$

this assumption could be expressed in a more precise way: each transformation of the Lorentz general group is a product of two "space" affine orthogonal transformations of Lorentz's transformation between time and a space coordinates. In ordinary terms and interpreting the manifold M_3 with the help of Decartes coordinates, we could assert that each transformation of Lorentz group is a set of two transformed coordinates and one motion with constant velocity. In such sense this formulation will be given in the second part of the book.

We will write the transformations denoted by the above mentioned symbols:

$$L\begin{pmatrix} x, & y, & z, & t \\ \overline{\overline{\overline{x}}}, & \overline{\overline{\overline{y}}}, & \overline{\overline{\overline{z}}}, & \overline{t} \end{pmatrix}:\quad \begin{aligned} \overline{\overline{\overline{x}}} &= \alpha + \alpha_1 x + \alpha_2 y + \alpha_3 z + \alpha_4 t, \\ \overline{\overline{\overline{y}}} &= \beta + \beta_1 x + \beta_2 y + \beta_3 z + \beta_4 t, \\ \overline{\overline{\overline{z}}} &= \gamma + \gamma_1 x + \gamma_2 y + \gamma_3 z + \gamma_4 t, \\ \overline{\overline{\overline{t}}} &= \delta + \delta_1 x + \delta_2 y + \delta_3 z + \delta_4 t, \end{aligned}$$

$$O\begin{pmatrix} x, & y, & z \\ \overline{x}, & \overline{y}, & \overline{z} \end{pmatrix}:\quad \begin{aligned} \overline{x} &= \mathfrak{a}_1 x + \mathfrak{a}_2 y + \mathfrak{a}_3 z, \\ \overline{y} &= \mathfrak{b}_1 x + \mathfrak{b}_2 y + \mathfrak{b}_3 z, \\ \overline{z} &= \mathfrak{c}_1 x + \mathfrak{c}_2 y + \mathfrak{c}_3 z, \\ \overline{t} &= t, \end{aligned}$$

$$\text{Л}\begin{pmatrix} \overline{x}, & \overline{t} \\ \overline{\overline{x}}, & \overline{t} \end{pmatrix}:\quad \begin{aligned} \overline{\overline{x}} &= a_1 \overline{x} + a_2 \overline{t}, \\ \overline{\overline{y}} &= \overline{y}, \\ \overline{\overline{z}} &= \overline{z}, \\ \overline{\overline{t}} &= b + b_1 \overline{x} + b_2 \overline{t}, \end{aligned}$$

$$O'\begin{pmatrix} x, & y, & z, & t \\ \bar{\bar{\bar{x}}}, & \bar{\bar{\bar{y}}}, & \bar{\bar{\bar{z}}}, & \bar{\bar{\bar{t}}} \end{pmatrix} : \quad \begin{aligned} \bar{\bar{\bar{x}}} &= \bar{a} + \bar{a}_1\bar{\bar{x}} + \bar{a}_2\bar{\bar{y}} + \bar{a}_3\bar{\bar{z}}, \\ \bar{\bar{\bar{y}}} &= \bar{b} + \bar{b}_1\bar{\bar{x}} + \bar{b}_2\bar{\bar{y}} + \bar{b}_3\bar{\bar{z}}, \\ \bar{\bar{\bar{z}}} &= \bar{c} + \bar{c}_1\bar{\bar{x}} + \bar{c}_2\bar{\bar{y}} + \bar{c}_3\bar{\bar{z}}, \\ \bar{\bar{\bar{t}}} &= t. \end{aligned}$$

In order that the product of the last three transformations gives the first transformation it is necessary and sufficient the following; the coefficient of these three transformations satisfies the bellow written equations:

$$\alpha = \mathfrak{a} \quad \beta = \bar{\mathfrak{b}} \quad \gamma = \bar{\mathfrak{c}} \quad \delta = b,$$

$$\begin{aligned} \alpha_1 &= a_1\mathfrak{a}_1\bar{\mathfrak{a}}_1 + \mathfrak{b}_1\bar{\mathfrak{a}}_2 + \mathfrak{c}_1\bar{\mathfrak{a}}_3, \\ \beta_1 &= a_1\mathfrak{a}_1\bar{\mathfrak{b}}_1 + \mathfrak{b}_1\bar{\mathfrak{b}}_2 + \mathfrak{c}_1\bar{\mathfrak{b}}_3, \\ \gamma_1 &= a_1\mathfrak{a}_1\bar{\mathfrak{c}}_1 + \mathfrak{b}_1\bar{\mathfrak{c}}_2 + \mathfrak{c}_1\bar{\mathfrak{c}}_3, \\ \alpha_2 &= a_1\mathfrak{a}_2\bar{\mathfrak{a}}_1 + \mathfrak{b}_2\bar{\mathfrak{a}}_2 + \mathfrak{c}_2\bar{\mathfrak{a}}_3, \\ \beta_2 &= a_1\mathfrak{a}_2\bar{\mathfrak{b}}_1 + \mathfrak{b}_2\bar{\mathfrak{b}}_2 + \mathfrak{c}_2\bar{\mathfrak{b}}_3, \\ \gamma_2 &= a_1\mathfrak{a}_2\bar{\mathfrak{c}}_1 + \mathfrak{b}_2\bar{\mathfrak{c}}_2 + \mathfrak{c}_2\bar{\mathfrak{c}}_3, \\ \alpha_3 &= a_1\mathfrak{a}_3\bar{\mathfrak{a}}_1 + \mathfrak{b}_3\bar{\mathfrak{a}}_2 + \mathfrak{c}_3\bar{\mathfrak{a}}_3, \\ \beta_3 &= a_1\mathfrak{a}_3\bar{\mathfrak{b}}_1 + \mathfrak{b}_3\bar{\mathfrak{b}}_2 + \mathfrak{c}_3\bar{\mathfrak{b}}_3, \\ \gamma_3 &= a_1\mathfrak{a}_3\bar{\mathfrak{c}}_1 + \mathfrak{b}_3\bar{\mathfrak{c}}_2 + \mathfrak{c}_3\bar{\mathfrak{c}},3 \end{aligned} \qquad (**)$$

$$\alpha_4 = a_2\bar{\mathfrak{a}}_1, \quad \beta_4 = a_2\bar{\mathfrak{b}}_1, \quad \gamma_4 = a_2\bar{\mathfrak{c}}_1,$$

$$\delta_1 = b_1\mathfrak{a}_1, \quad \delta_2 = b_1\mathfrak{a}_2, \quad \delta_3 = b_1\mathfrak{a}_3$$

$$\delta_4 = b_2.$$

If we assume that we have twenty quantities $\alpha_1, \beta, \ldots$ given to us arbitrary, than the system of 20 equations we get can be used for defining the quantities: $\mathfrak{a}_1, \mathfrak{a}_2, \ldots \mathfrak{a}_1, \mathfrak{a}_2, \ldots \bar{\mathfrak{a}}, \bar{\mathfrak{a}}_1, \bar{\mathfrak{a}}_2, \ldots$; otherwise, for defining $9 + 5 + 12 = 26$ quantities; it is obvious that part of these 26 quantities satisfying the equations $(**)$, will be arbitrary; let us assume, that the transformation $O'\left(\frac{\bar{\bar{x}}, \bar{\bar{y}}, \bar{\bar{z}}}{\bar{\bar{\bar{x}}}, \bar{\bar{\bar{y}}}, \bar{\bar{\bar{z}}}}\right)$ will be orthogonal affine transformation, then on the 26 quantities will be imposed 6 restricting

90

them equations, namely the equations:

$$\bar{a}_1^2 + \bar{b}_1^2 + \bar{c}_1^2 = 1,$$
$$\bar{a}_2^2 + \bar{b}_2^2 + \bar{c}_2^2 = 1,$$
$$\bar{a}_3^2 + \bar{b}_3^2 + \bar{c}_3^2 = 1,$$
$$\bar{a}_2\bar{a}_3 + \bar{b}_2\bar{b}_3 + \bar{c}_2\bar{c}_3 = 0,$$
$$\bar{a}_3\bar{a}_1 + \bar{b}_3\bar{b}_1 + \bar{c}_3\bar{c}_1 = 0,$$
$$\bar{a}_1\bar{a}_2 + \bar{b}_1\bar{b}_2 + \bar{c}_1\bar{c}_2 = 0.$$

$$(\ast\ast\ast)$$

We will show that when $\alpha, \beta, \alpha_1, \beta_1 \ldots$, satisfying the conditions of pseudo-orthogonality, i.e, the equations $(\ast)$, always can be found such $\bar{a}, \bar{b}, \bar{a}_1, \bar{b}_1, \ldots, a_1, b_1, \ldots, a_1, b_1, \ldots$, in order that both systems of equations $(\ast\ast)$ and $(\ast\ast\ast)$ are satisfied.

Multiplying the equations for $\alpha_1, \beta_1, \gamma_1$, in the system $(\ast\ast)$, by $\bar{a}_1, \bar{b}_1, \bar{c}_1$ and adding the obtained equations, we find, taking into account the conditions $(\ast\ast\ast)$, the following equation:

$$a_1a_1 = \alpha_1\bar{a}_1 + \beta_1\bar{b}_1 + \gamma_1\bar{c}_1,$$

In the same way we can solve the equations $(\ast\ast)$, with respect to $b_1, c_1, a_1a_2, b_2, a_1a_3, b_3, c_3$. Carrying out the above mentioned transformation we will rewrite the equations $(\ast\ast)$ in the following way:

$$\bar{a} = \alpha, \quad \bar{b} = \beta, \quad \bar{c} = \gamma, \quad b = \delta,$$

$$a_1a_1 = \alpha_1\bar{a}_1 + \beta_1\bar{b}_1 + \gamma_1\bar{c}_1,$$
$$b_1 = \alpha_1\bar{a}_2 + \beta_1\bar{b}_2 + \gamma_1\bar{c}_2,$$
$$c_1 = \alpha_2\bar{a}_3 + \beta_1\bar{b}_3 + \gamma_1\bar{c}_3,$$
$$a_1a_2 = \alpha_2\bar{a}_1 + \beta_2\bar{b}_1 + \gamma_2\bar{c}_1,$$
$$b_2 = \alpha_2\bar{a}_2 + \beta_2\bar{b}_2 + \gamma_2\bar{c}_2,$$
$$c_2 = \alpha_2\bar{a}_3 + \beta_2\bar{b}_3 + \gamma_3\bar{c}_3,$$
$$a_1a_3 = \alpha_3\bar{a}_1 + \beta_3\bar{b}_1 + \gamma_3\bar{c}_1,$$
$$b_3 = \alpha_3\bar{a}_2 + \beta_3\bar{b}_2 + \gamma_3\bar{c}_2,$$
$$c_3 = \alpha_3\bar{a}_3 + \beta_3\bar{b}_3 + \gamma_3\bar{c}_3,$$
$$a_2^2 = \alpha_4^2 + \beta_4^2 + \gamma_4^2,$$

$$a_2\bar{a}_1 = \alpha_4, \quad a_2\bar{b}_1 = \beta_4, \quad a_2\bar{c}_1 = \gamma_4 \quad b_2 = \delta_4,$$

$$b_1a_1 = \delta_1 \quad b_1a_2 = \delta_2, \quad b_1a_3 = \delta_3.$$

The first $4+9 = 13$ equations show, that by the given $\alpha, \beta, \gamma, \alpha_1, \beta_1, \ldots$ and by $\bar{a}_1, \bar{a}_2, \ldots$ satisfying the conditions $(\ast\ast)$, we could define the quantities: $\bar{a}, \bar{b}, \bar{c}$, b, a_1a_1, b_1, c_1, a_1a_2, $b_2, c_2, a_1a_3, b_3, c_3$; fourteenth and the three following equations show that we can always choose a_2 and two of the quantities $\bar{a}_1, \bar{b}_1, \bar{c}_1$ satisfying the condition $(***)$, for these for equations.

I fact, the system of the nine quantities $\bar{a}_1, \bar{a}_2, \ldots$ is connected to the six relations of the orthogonality and has three arbitrary parameters; we can choose two from these three parameters in such a way that to give the quantities a_1, b_1, c_1, any values, satisfying the condition $\bar{a}_1^2 + \bar{b}_1^2 + \bar{c}_1^2 = 1$; if $\alpha_4 = \beta_4 = \gamma_4 = 0$, then the obtained equations from fourteenth to seventeenth will be satisfied if $\alpha_2 = 0$; if even one of $\alpha_4, \beta_4, \gamma_4$ will be different from zero, then a_2 will not be $= 0$ and with the help of our equations we will get the next equations for $\bar{a}_1, \bar{b}_1, \bar{c}_1$:

$$\bar{a}_1 = \frac{\alpha_4}{a_2}, \quad \bar{b}_1 = \frac{\beta_4}{a_2}, \quad \bar{c}_1 = \frac{\gamma_4}{a_2}$$

taking into account that $\bar{a}_1^2 + \bar{b}_1^2 + \bar{c}_1^2 = 1$; according to the note made above we can always choose two out of three arbitrary parameters belonging to the system of nine values $\bar{a}_1, \bar{a}_2, \ldots$ satisfying the conditions $(***)$, in such way that the written equations would have been satisfied.

There are left to discuss only the last four equations: the eighteenth of the equations $(**)$ will define b_2, concerning the last three equations, they will be placed in the relations $(*)$ of pseudo-orthogonality as far as one of these equations will be satisfied.

In order that we clarify this, we will discuss the last three out of $(*)$ relations, value δ_4 could not be a zero, because $\delta_4 = 0$ the fourth of the relations $(*)$ could be rewritten as: $(\alpha_4)^2 + (\beta_4)^2 + (\gamma_4)^2 = -c^2$, which is obviously impossible, because the sum of the three quadratic real numbers could not be a negative number; because c is also different from zero, then the last three of the relations $(*)$ can be solved with respect to $\delta_1, \delta_2, \delta_3$ and then they will be written in the following way;

$$\delta_1 = \frac{\alpha_1\alpha_1 + \beta_1\beta_4 - \gamma_1\gamma_4}{c^2\delta_4},$$

$$\delta_2 = \frac{\alpha_2\alpha_4 + \beta_2\beta_4 + \gamma_2\gamma_4}{c^2\delta_4},$$

$$\delta_3 = \frac{\alpha_1\alpha_4 + \beta_3\beta_4 + \gamma_3\gamma_4}{c^2\delta_4}$$

Using those of the relations $(**)$ which do not include $\delta_1, \delta_2, \delta_4$ and taking into account the conditions $(* * *)$, we find:

$$\delta_1 = \frac{a_1}{c^2 b_2}\, a_1, \quad \delta_2 = \frac{a_1}{c^2 b_2}\, a_2, \quad \delta_3 = \frac{a_1}{c^2 b_4}\, a_3$$

From here it is seen that from the condition $(*)$ of pseudo-orthogonality one can write the equation:

$$\frac{\delta_1}{c^2 b_2} = \frac{\delta_1}{a_2} = \frac{\delta_2}{a_3},$$

i.e. from the last three equations $(**)$ it is sufficient that one of them had been satisfied in order that the remaining two also hold due to the condition $(*)$. For this only equation we have several values, including the third arbitrary parameter of the orthogonal system of values namely $\bar{a}_1, \bar{a}_2 \dots$. By choosing properly this parameter, we will satisfy the last from the relations $(**)$, while the conditions $(***)$ will hold. Except this, the values a_1 and b_1 will be arbitrary, and we can chose them through the condition: $(****)$

$$(a_1)^2 - c^2 (b_1)^2 = 1. \tag{****}$$

It follows from the previous considerations that we can find $\bar{a}, \bar{b}$, $\bar{a}_1, \bar{b}_1, \dots, a_1, b_1, \dots a_1, b_1, \dots$ in order that the conditions of pseudo-orthogonality $(*)$, satisfy the equations $(**)$, where the conditions $(***)$ and $(****)$ hold.

Let us show now that the orthogonality of the transformation O and the belonging of the transformation Π to the Lorentz group follow from the pseudo-orthogonal conditions.

Using the first seven equations of $(*)$ and taking into account the relations $(**)$ and $(***)$ we find:

$$a_1^2 \mathfrak{a}_1^2 + \mathfrak{b}_1^2 + \mathfrak{c}_1^2 - c^2 b_1^2 \mathfrak{a}_1^2 = 1,$$
$$a_1^2 \mathfrak{a}_2^2 + \mathfrak{b}_2^2 + \mathfrak{c}_2^2 - c^2 b_1^2 \mathfrak{a}_2^2 = 1,$$
$$a_1^2 \mathfrak{a}_3^2 + \mathfrak{b}_3^2 + \mathfrak{c}_3^2 - c^2 b_1^2 \mathfrak{a}_3^2 = 1,$$

$$a_2^2 - c^2 b_2^2 = -c^2,$$

$$a_1^2 \mathfrak{a}_2 \mathfrak{a}_3 + \mathfrak{b}_2 \mathfrak{b}_3 + \mathfrak{c}_2 \mathfrak{c}_3 - c^2 b_1^2 \mathfrak{a}_2 \mathfrak{a}_3 = 0,$$
$$a_1^2 \mathfrak{a}_3 \mathfrak{a}_1 + \mathfrak{b}_3 \mathfrak{b}_1 + \mathfrak{c}_3 \mathfrak{c}_1 - c^2 b_1^2 \mathfrak{a}_3 \mathfrak{a}_1 = 0,$$
$$a_1^2 \mathfrak{a}_1 \mathfrak{a}_2 + \mathfrak{b}_1 \mathfrak{b}_2 + \mathfrak{c}_1 \mathfrak{c}_2 - c^2 b_1^2 \mathfrak{a}_1 \mathfrak{a}_2 = 0.$$

From these equations we find by taking into account that $(a_1)^2 - c^2(b_1)^2 = 1$:

$$a_1^2 + b_1^2 + c_1^2 = 1,$$
$$a_2^2 + b_2^2 + c_2^2 = 1,$$
$$a_3^2 + b_3^2 + c_3^2 = 1,$$
$$a_2 a_3 + b_2 b_3 + c_2 c_3 = 0,$$
$$a_3 a_1 + b_3 b_1 + c_3 c_1 = 0,$$
$$a_1 a_2 + b_1 b_2 + c_1 c_2 = 0,$$

which will prove that the transformations O belong to the affine orthogonal transformations in M_3.

In order that we clarify the character of the transformation Л, we use the last three equations from (*) and rewrite them in the following way:

$$(a_1 a_2 - c^2 b_1 b_2)a_1 = 0,$$
$$(a_1 a_2 - c^2 b_1 b_2)a_2 = 0,$$
$$(a_1 a_2 - c^2 b_1 b_2)a_3 = 0,$$

As $a_1^2 + a_2^2 + a_3^2 = 1$, then a_1, a_2, a_3 cannot vanish simultaneously, because from the previous equations we will have:

$$a_1 a_2 - c^2 b_1 b_2 = 0,$$

i.e. a_1, a_2, b_1, b_2 will satisfy the relations:

$$(a_1)^2 - c^2(b_1)^2 = 1,$$
$$a_1 a_2 - c^2 b_1 b_2 = 0,$$
$$(a_2)^2 - c^2(b_2)^2 = -c^2,$$

i.e. the transformation Л belonsg to the general Lorentz group in M_2.

The above discussion, demonstrating that each general Lorentz transformation in M_4 reduces to two orthogonal transformations in M_3 and to the Lorentz transformation in M_2, shows that the most important element in the general Lorentz transformations in M_4 is the Lorentz transformation in M_2 which we studied above.

The limiting case of expanded Lorentz transformation when $c = \infty$

94

will be a transformation, expressed by the equations:

$$\bar{x} = \alpha + \alpha_1 x + \alpha_2 y + \alpha_3 z + \alpha_4 t,$$
$$\bar{y} = \beta + \beta_1 x + \beta_2 y + \beta_3 z + \beta_4 t,$$
$$\bar{z} = \gamma + \gamma_1 x + \gamma_2 y + \gamma_3 z + \gamma_4 t,$$
$$\bar{t} = \delta + \delta_4 t.$$

A similar kind of transformations form, as it is easy to check by direct calculations, a group of transformation $\boxed{G}_{14}$, which is completely analogous to the Galilean group in M_2; this group will be called *expanded Galilean group*; it has the following invariant equation:

$$(t^{(2)} - t^{(1)})^2 = 0.$$

A general Galilean group will be called such a subgroup $\boxed{G}_{13}$ of the expanded Galilean group, in which $\delta_4 = 1$; the invariant of this general Galilean group is the expressen:

$$J = (t^{(2)} - t^{(1)})^2.$$

If t is a time coordinate, then the Galilean group completely distinguishes this time coordinate from the space coordinates. Although the Lorentz group of transformations assigns special properties to the time coordinate, it does not distinguish it from the space coordinates so sharply; the real meaning of this remark will be seen in the chapter devoted to the special principle of relativity.

In conclusion of this paragraph we notice that all of our considerations are significantly simplified if, instead of a real coordinate t, we introduce a purely imaginary number l linking it with t by the relation:

$$l = cit,$$

The pseudo-orthogonal transformations in M_4, with coordinates (x, y, z, t) will go formally in the orthogonal transformation in a "manifolds" with elements defined by the quantities (x, y, z, l) where the value l will be a purely imaginary number.

As so far we were operating exclusively with manifolds, which elements have been defined by *real* numbers, then when we discussed the group of the Lorentz transformations we did not use the complex quantity l. The specificity of the time coordinate, clearly manifested when we do not use the imaginary quantity, but it is suppressed when we introduce the imaginary quantity. It is this reason and the intention to reveal fully the idea of manifolds with elements, defined only by real numbers, for not using in this paragraph the brilliant idea (belonging to Minkowski) of introducing the imaginary coordinate l instead of a real time coordinate.

3   Tensor Algebra

1. Having clarified the notion of tensor in the previous section, we will turn to algebraic operations with tensors. We will use the ordinary algebraic symbolism or, as Weyl notices rightfully in his book "Raum Zeit, Materie", trying to introduce the special symbolism for tensorial algebra and analysis, in most cases only obscures the issue.

Addition and multiplication of tensors by a number is done in such a way that in the first case by adding the components of the tensors which we add we obtain the components of a new tensor, whereas in the second case the components of the tensor are all multiplied by a given number. There is no need to state specifically that both operations with the tensors will give again tensors.

For example: $\mathbf{T}_{ik} + \mathbf{Q}_{ik}$ is a tensor with components $T_{ik} + Q_{ik}$. In order to prove that the obtained product is a tensor, multiply its common component by R^{ik}, where R^{ik} is an arbitrary cogradient tensor

$$(T_{ik} + Q_{ik})\, R^{ik} = T_{ik} R^{ik} + Q_{ik} R^{ik}$$

which is a sum of scalars, i.e. a scalar, so by theorem 4, $\mathbf{T}_{ik} + \mathbf{Q}_{ik}$ is a contragradient tensor.

It is more complicated to obtain the product of tensors; like in three dimensional vector analysis we distinguish two kinds of products: external – producing a vector, and internal – producing a scalar, similarly in tensor algebra we distinguish the *product of tensors*, increasing or leaving without change the rank of the tensors that are multiplied and *the composition of tensors* – an operation decreasing the rank of the tensors which are participating in these operations.

The product of tensors

$$\mathbf{T}^{i_1 i_2 \ldots i_r}_{j_1 j_2 \ldots j_s}$$

and

$$\mathbf{Q}^{k_1 k_2 \ldots k_\rho}_{l_1 l_2 \ldots l_\sigma}$$

is called the tensor

$$\mathbf{R}^{i_1 i_2 \ldots i_r \quad k_1 k_2 \ldots k_\rho}_{j_1 j_2 \ldots j_s \quad l_1 l_2 \ldots l_\sigma}$$

whose components are defined by the equations:

$$\mathbf{R}^{i_1 i_2 \ldots i_r \quad k_1 k_2 \ldots k_\rho}_{j_1 j_2 \ldots j_s \quad l_1 l_2 \ldots l_\sigma} = \mathbf{T}^{i_1 i_2 \ldots i_r}_{j_1 j_2 \ldots j_s} \, \mathbf{Q}^{k_1 k_2 \ldots k_\rho}_{l_1 l_2 \ldots l_\sigma}.$$

Multiplying two cogradiant vectors, we obtain a cogradiednt tensor of the second rank, multiplying two contragradient vectors we obtain a contragredient tensor of the second rank and so on:

$$T^{il} = a^i b^k, \quad T_{ik} = a_i b_k, \quad T^k_i = a_i b^k, \quad T^{kl}_i = a_i b^k c^l.$$

A question may arise of whether each tensor is formed by the product of tensors of low rank or it is not the case. Let us clarify this question for the contragradient tensor and for simplicity we assume that $n = 3$; then the 9 components (3^2) of the contragredient tensor will be: T_{ik}, ($i, k = 1, 2, 3,$); to obtain this tensor through the product of the vectors $\mathbf{a}_i$ and $\mathbf{b}_k$ the following condition should be satisfied:

$$a_i = \frac{T_{ik}}{b_k}, \qquad (i = 1, 2, 3, \quad k = 1, 2, 3).$$

Therefore, for defining b_1, b_2, b_3 we will have six equations:

$$\frac{T_{i1}}{b_1} = \frac{T_{i2}}{b_2} = \frac{T_{i3}}{b_3}, \quad (i = 1, 2, 3,).$$

From here it is clear that it is impossible, in the general case, to satisfy the six equations of the three quantites; b_1, b_2, b_3.

It is not difficult to see, that each tensor T_{ik} can be formed as a sum of n tensors of the type $a_i b_k$, but we will not discuss this question.

2. In many cases of the applications of tensor calculus an important role is played by the property of symmetry, which is typical for some tensors. *A tensor is symmetric with respect to any two of its indices, if the permutation of these indices do not change the value of the corresponding component. A tensor is antisymmetric with respect to any two of its indices, if the permutation of these indices is changing the sign of the corresponding component, but does not change the numeric value of the component.*

For example, contragradient tensor $\mathbf{T}_{ik}$ is symmetric with respect to its indices, if for all i, k the equation $T_{ik} = T_{ki}$ holds; the fundamental tensor should, by definition, be a symmetric tensor. It is not difficult to prove that the conjugated fundamental tensor will be a cogradiant symmetric tensor.

The property of symmetry cannot be applied to all indices; for example a tensor of the third rank $\mathbf{T}_{ikl}$, which is symmetric with respect to the first two indices, satisfies for all values of the indices the equation: $\mathbf{T}_{ikl} = \mathbf{T}_{kil}$. However, a permutation of the third index can change completely the value of the corresponding component of the tensor.

The property of symmetry of a tensor decreases the number of its different components; the symmetric tensor of the second rank will have not n^2, but $\frac{n(n+1)}{2}$ components. In such a way, the fundamental tensor will have $\frac{n(n+1)}{2}$ components, i.e., for $n = 2$, 3 components, for $n = 3$, 6, for $n = 4$, 10 components. The number of the components of the symmetric tensor of the second rank is defined by the following rule: the equation $T_{ik} = T_{ki}$ becomes identity for $k = i$. In such a way, out of n^2 components of the tensor n components of T_{ii} do not satisfy any relations; the remaining $n^2 - n$ components are pairwise equal (due to symmetry). For this reason, among them different values will have $\frac{n^2-n}{2}$, we will have $n + \frac{n^2-n}{2} = \frac{n(n+1)}{2}$ different components of the symmetric tensor of the second rank.

The antisymmetric contragradient tensor of the second rank satisfies the relation: $T_{ik} = -T_{ki}$. From this relation it follows that $T_{ik} = 0$. In such a way an antisymmetric tensor of the second rank has only $\frac{n(n-1)}{2}$ completely different components; for $n = 3$, we have 3 components, so each antisymmetric tensor of the second rank in M_3 can be related to a tree-dimensional vector. We have seen above, that the vector product of two three-dimensional vectors, strictly speaking, cannot be considered as a vector. Later we will see that three components of this vector product can be viewed as three completely different components of some antisymmetric tensor of the second rank.

In order to form examples of symmetric and antisymmetric tensors, we find the product of two vectors and form two tensors:

$$S_{ik} = a_i b_k + a_k b_i, \quad T_{ik} = a_i b_k - a_k b_i$$

The first of these tensors is obviously a symmetric tensor, the second is an antisymmetric tensor.

We choose two three-dimensional vectors $\mathfrak{U}$ and $\mathfrak{B}$, denote them, according to the general-tensorial symbols, by $\mathbf{a}_i$ and $\mathbf{b}_k$ and form in M_3 a tensor $T_{ik} = a_i b_k - a_k b_i$:

$$T_{11} = 0, \quad T_{12} = a_1 b_2 - a_2 b_1, \quad T_{13} = a_1 b_3 - a_3 b_1,$$

$$T_{21} = a_2 b_1 - a_1 b_2, \quad T_{22} = 0, \quad T_{23} = a_2 b_3 - a_3 b_2$$

$$T_{21} = a_3 b_1 - a_1 b_3 \quad T_{32} = a_3 b_2 - a_2 b_3, \quad T_{23} = 0$$

The formed tensor will be an antisymmetric tensor and its three completely different components can be related to the three components of the vector product of the vectors $\mathfrak{U}$ and $\mathfrak{B}$ by the following formulas:

$$\mathfrak{G} = \left[\mathfrak{U}, \mathfrak{B} \right]$$

$$T_{23} = -T_{22} = \mathfrak{G}_1, \quad T_{31} = -T_{13} = \mathfrak{G}_2, \quad T_{12} = -T_{21} = \mathfrak{G}_3$$

In addition to symmetry and antisymmetry we have also *cyclic symmetry.*

A tensor possesses cyclic symmetry in any three of its indexes, if the sum of the components of the tensor, formed by a cyclic permutation of these three indices is always equal to zero.

A contragradient tensor of the third rank $\mathbf{T}_{ikl}$ possesses a cyclic symmetry if for all values of the indecis i, k, l the following equation holds:

$$T_{ikl} + T_{kli} + T_{lik} = 0.$$

When we study tensor analysis we will give examples of tensors, having cyclic symmetry.

If $\mathbf{a}_i, \mathbf{b}_k, \mathbf{c}_l$ are contragradient vectors, then it is not difficult to see, that the tensor T_{ikl} whose components are defined by the equations

$$T_{ikl} = a_i b_k c_l - a_k b_l c_i$$

possesses the property of cyclic symmetry.

Indeed:

$$T_{ikl} = a_i b_k c_l - a_k b_l c_i; \quad T_{kli} = a_k b_l c_i - a_l b_i c_k, \quad T_{lik} = a_l b_i c_k - a_i b_k c_l$$

means

$$T_{ikl} + T_{kli} + T_{lik} = 0$$

It is not difficult, that the property of symmetry, antisymmetry and cyclic symmetry of tensors do not change under the transformation of variables and are properties, so to speak, invariant with respect to the transformation of the variables.

Let us prove, that the property of symmetry of the tensor $\mathbf{T}_{ik}$ implies the symmetry of the tensor $\overline{\mathbf{T}}_{ik}$:

$$\overline{T}_{ik} = T_{\alpha\beta} \frac{\partial x_\alpha}{\partial \overline{x}_i} \frac{\partial x_\beta}{\partial \overline{x}_k}, \quad \overline{T}_{ki} = T_{\alpha\beta} \frac{\partial x_\alpha}{\partial \overline{x}_k} \frac{\partial c_\beta}{\partial \overline{x}_i} = T_{\beta\alpha} \frac{\partial x_\alpha}{\partial \overline{x}_i} \frac{\partial x_\beta}{\partial \overline{x}_k},$$

but $T_{\beta\alpha} = T_{\alpha\beta}$ because $\overline{T}_{ki} = \overline{T}_{ik}.$

In a completely analogous way the invariance of the anti and cyclic symmetry of tensors is proved. In such a way the property of symmetry is a property of the tensors and does not depend on the choice of variables for forming the manifold M_n.

In the same way we can prove the following property of the contragradient (or cogradient) tensor of the fourth rank $\mathbf{T}_{iklm}$. If in one of the manifolds M_n the components of the tensor $\mathbf{T}_{iklm}$ have one of the following properties of symmetry:

$$T_{iklm} = -T_{ikml}, \quad T_{iklm} = -T_{kilm},$$

$$T_{iklm} = T_{lmik},$$

$$T_{iklm} + T_{ilmk} + T_{imkl} = 0$$

then in any other manifold $\overline{M}_n$ the components of our tensor have the same properties of symmetry. We will need this property of tensors when derive the equations of electrodynamics.

3. When we consider the operations of the compositions of tensors, generally speaking, operations, decreasing the rank of a tensor, it is necessary to rely on theorem 6 proved above; with the help of this theorem, the operation of the *composition* of tensors will be defined in the following way:

A composition of two tensors $\mathbf{T}^{\alpha_1\alpha_2...\alpha_p\ k_1k_2...k_\rho}_{\beta_1\beta_2...\beta_q\ l_1l_2...l_\sigma}$ *and* $\mathbf{Q}^{\beta_1\beta_2...\beta_q\ i_1i_2...i_r}_{\alpha_1\alpha_2...\alpha_p\ j_1j_2...j_s}$ *is called the tensor* $\mathbf{R}^{i_1i_2...i_r\ k_1k_2...k_\rho}_{j_1j_2...j_s\ l_1l_2...l_\sigma}$ *whose components are defined by the formulas:*

$$R^{i_1i_2...i_r\ k_1k_2...k_\rho}_{j_1j_2...j_r\ l_1l_2...l_\sigma} = T^{\alpha_1\alpha_2...\alpha_p\ k_1k_2...k_\rho}_{\beta_1\beta_2...\beta_q\ l_1l_2...l_\sigma}\ Q^{\beta_1\beta_2...\beta_q\ i_1i_2...i_r}_{\alpha_1\alpha_2...\alpha_p\ j_1j_2...j_s}.$$

In order to clarify this definition let us have a composition of the tensor $\mathbf{T}^l_{ik}$ with the tensor $\mathbf{Q}^m_j$. We can produce a composition by different methods obtaining different tensors:

$$R^m_{ik} = T^\alpha_{ik}\,Q^m_\alpha, \quad P^l_{km} = T^l_{\alpha k}\,Q^\alpha_m, \quad R_i = T^\alpha_{i\beta}\,Q^\beta_\alpha,$$

In some cases the rank of the obtained tensors is not less than the rank of the the tensors that form the composition. In other cases the rank is greater. In most cases the aim of the composition of tensors is to lower their rank. *Einstein* calls this process "contraction" – Verjüngung – of tensors; such kind of a process has its natural limit, when the composition of tensors results in a scalar.

They exist two kinds of compositions, playing a specific role in tensor calculus; the first type of composition is formed, if the unit tensor is takes as one of the tensors in the composition; we will see that the result of this composition almost always will be lowering the rank of the companion tensor, in other words, removing several indeces from the tensor. We will call such a composition *unit composition*; its result is either removing two indices from the companion tensor (not the unit tensor) or leaving this tensor without changes.

In fact we have:

$$T^{i_1 i_2 \ldots i_r}_{\alpha j_2 j_3 \ldots j_s} \, \delta^{\alpha}_{j_1} = T^{i_1 i_2 \ldots i_r}_{j_1 j_2 \ldots j_s},$$

$$T^{\alpha i_2, i_3 \ldots i_r}_{j_1 j_1 j_2 \ldots j_s} \, \delta^{i_1}_{\alpha} = T^{i_1 i_2 \ldots i_r}_{j_1 j_2 \ldots j_s}.$$

In other words, the companion tensor was left unchanged.

On the other hand, double sums will give:

$$\mathbf{T}^{\beta i_2 i_3 \ldots i_r}_{\alpha j_2 j_3 \ldots j_s} \, \delta^{\alpha}_{\beta} = \mathbf{T}^{\alpha i_2 i_3 \ldots i_r}_{\alpha j_2 j_3 \ldots j_s} = \sum_{\alpha=1}^{n} \mathbf{T}^{\alpha i_2 i_3 \ldots i_r}_{\alpha j_2 j_3 \ldots j_s} = T^{i_2 j_3 \ldots i_r}_{j_2 j_3 \ldots j_s}.$$

So, we got a tensor, in which two indices i_1 and j_1 were eliminated.

Let us give an example how from the tensor of the fourth rank $\mathbf{F}^{m}_{ilk}$ we can form, with the help of the unit composition a tensor of the second rank:

$$F_{ik} = F^{\beta}_{i \alpha k} \, \delta^{\alpha}_{\beta} = F^{\alpha}_{i \alpha k} = \sum_{\alpha=1}^{n} F^{\alpha}_{i \alpha k}$$

While the first kind of composition – the unit composition allows lowering the rank of a tensor by eliminating indices in it – the second kind of composition makes it possible to change in the tensor the place of indices from down to up and vice-verse, in other words raise and lower indices, i.e. to juggle with them, using the figurative expression of Weyl. This kind of composition is often used when as one of the companion tensors is the fundamental or the conjugated fundamental tensor ($\mathbf{g}_{ik}$ or ($\mathbf{g}^{ik}$) and is called *principal composition*.

Doing the principal composition on the tensor $\mathbf{T}^{i_1, i_2 \ldots i_r}_{j_1 j_2 \ldots j_s}$ we find:

$$T^{i_1 i_2 \ldots i_{r-1} \alpha}_{j_1 j_2} \ldots j_s \, g_{\alpha j_{s+1}} = T^{i_1 i_2 \ldots i_{r-1}}_{j_1 j_2 \ldots j_s j_{s+1}},$$

$$T^{i_1 i_2 \ldots i_r}_{j_1 j_2 \ldots j_{s-1} \alpha} \, g^{\alpha i}_{r+1} = T^{j_1 i_2 \ldots i_r i_{r+1}}_{i_1 j_2 \ldots j_{s-1}}$$

In the first case we have lowering of the indices, going from from cogradient to contragradient; in the second case – raising the indices – going from contragradient into cogradient.

Let us give several examples of similar "juggling" with the indices:

$$a_k = a^{\alpha} g_{\alpha} k, \quad a^l = a_{\alpha} g^{\alpha i}$$

$$a_{ik} = a_i^\alpha g_{\alpha k}, \quad a_i^k = a_{i\alpha} g^{\alpha k}, \quad a^{ik} = a_\alpha^i g^{\alpha k}$$

Later we will denote, if is possible, the tensor, obtained through the principal composition from a given tensor, by the same letter while changing only (according to the principal composition) the arrangement of indices.

It is not difficult to see, that the principal composition applied to the same tensor, first to raise it, second to lower the indices, does not change the tensor at all.

In fact:

$$T^{i_1 i_2 \ldots i_{r-1}}_{j_1 j_2 \ldots j_s j_{s+1}} = T^{i_1 i_2 \ldots i_{r-1} \alpha}_{j_1 j_2 \ldots j_s} g_{\alpha j_{s+1}}$$

by raising the indices and denoting the obtained tensor by(to avoid misunderstanding):

$$Q^{i_1 i_2 \ldots i_{r-1} i_r}_{j_1 j_2 \ldots j_s},$$

we get

$$Q^{i_1 i_2 \ldots i_r}_{j_1 j_2 \ldots j_s} = T^{i_1 i_2 \ldots i_{r-1}}_{j_1 j_2 \ldots j_s \beta} g^{\beta i_r} = T^{i_1 i_2 \ldots i_{r-1} \alpha}_{j_1 j_2 \ldots j_s} g_{\alpha \beta} g^{\beta i_r}$$

but $g_{\alpha\beta} g^{\beta i_r} = \delta^{i_r}\alpha$, for this reason

$$Q^{i_1 i_2 \ldots i_r}_{j_1 j_2 j_s} = T^{i_1 i_2 \ldots i_{r-1} \alpha}_{j_1 j_2 \ldots j_s} \delta^{i_r}_\alpha = T^{i_1 i_2 \ldots i_{r-1} i_r}_{j_1 j_2 \ldots j_3},$$

which proves our assertion.

4. In the conclusion of this section we will point out several more tensors, which will play an important role later, namely in electrodynamics. Taking into account further applications, we will confine ourselves to the manifold M_4 of fourth dimensions – in such a way the indices will take on values from one to four.

Considering any arrangement *iklm* of different indices $1, 2, 3, 4$ we can divide them into two classes.

We will call *even arrangements* such arrangements, which can be obtained from an even number of transpositions of the natural arrangement $1, 2, 3, 4$ (transposition is a permutation of any two indices).

Uneven arrangements will be the remaining arrangements obtained from the natural arrangement $1, 2, 3, 4$ by an uneven number of transpositions.

We willdenote by the symbol δ_{iklm} the quantity which turns into zero when any two of its indices are equal, and becomes $+1$ or in -1 depending on whether the arrangement of two different indices of *iklm* is even or uneven.

With the help of this symbol it is not difficult to write a determinant of the fourth order $||a_{ik}||$ in the following fourfold sum:

$$\delta_{iklm}\,||a_{\alpha\beta}|| = \begin{vmatrix} a_{1i} & a_{2i} & a_{3i} & a_{4i} \\ a_{1k} & a_{2k} & a_{3k} & a_{4k} \\ a_{1l} & a_{2l} & a_{3l} & a_{4l} \\ a_{1m} & a_{2m} & a_{3m} & a_{4m} \end{vmatrix} = (\delta_{\alpha\beta\gamma\delta}\,a_{\alpha i}\,a_{\beta k}\,a_{\gamma l}\,a_{\delta m})$$

In such way we will have

$$\delta_{iklm}\,D = \left(\delta_{\alpha\beta\gamma\delta}\,\frac{\partial x_\alpha}{\partial \overline{x}_i}\,\frac{\partial x_\beta}{\partial \overline{x}_k}\,\frac{\partial x_\gamma}{\partial \overline{x}_l}\,\frac{\partial x_\delta}{\partial \overline{x}_m}\right)$$

where,

$$D = \left\|\frac{\partial x_i}{\partial \overline{x}_k}\right\|.$$

Let us prove, that $\mathbf{G}_{iklm}$ defined by the equations:

$$G_{iklm} = \sqrt{g}\,\delta_{iklm} \qquad g = ||g_{\alpha\beta}||$$

is a contragradient tensor. In fact we have:

$$\overline{G}_{iklm} = \sqrt{\overline{g}}\,\delta_{iklm},$$

but

$$\sqrt{\overline{g}} = \sqrt{g}.D,$$

therefore:

$$\overline{G}_{iklm} = \sqrt{\overline{g}}\,\delta_{iklm}\,D = \sqrt{g}\,\delta_{\alpha\beta\gamma\delta}\,\frac{\partial x_\alpha}{\partial \overline{x}_i}\,\frac{\partial x_\beta}{\partial \overline{x}_k}\,\frac{\partial x_\gamma}{\partial \overline{x}_l}\,\frac{\partial x_\delta}{\partial \overline{x}_m} =$$

$$= G_{\alpha\beta\gamma\delta}\,\frac{\partial x_\alpha}{\partial \overline{x}_i}\,\frac{\partial x_\beta}{\partial \overline{x}_k}\,\frac{\partial x_\gamma}{\partial \overline{x}_l}\,\frac{\partial x_\delta}{\partial \overline{x}_m},$$

which proves the tensor nature of of $\mathbf{G}_{iklm}$. This tensor is called the tensor of Levi-Chivita.

Using the principal composition it is possible from the contragradient tensor of Levi-Chivita to form the following cogradient tensor G^{iklm}

$$G^{iklm} = G_{\alpha\beta\gamma\delta}\,g^{\alpha i}\,g^{\beta k}\,g^{\gamma l}\,g^{\delta m}.$$

We will prove that this tensor satisfies the following equation:

$$G^{iklm} = \frac{1}{\sqrt{g}}\,\delta_{iklm}.$$

In fact, by the definition of the considered tensor, we will have:

$$G^{iklm} = \sqrt{g}\, \delta_{\alpha\beta\gamma\delta}\, g^{\alpha i}\, g^{\beta k}\, g^{\gamma l}\, g^{\delta m} = \sqrt{g}\, \delta_{iklm} \left\| g^{\lambda m} \right\| =$$

$$= \frac{i}{g}\, \sqrt{g}\, \delta_{iklm} = \frac{1}{\sqrt{g}}\, \delta_{iklm},$$

which proves the above equation.

In the same way, using the principal composition, we can form the mixed tensor $\mathbf{G}_{ik}^{lm}$:

$$G_{ik}^{lm} = G_{ik\alpha\beta}\, g^{\alpha l}\, g^{\beta m} = \sqrt{g}\, \delta_{ik\alpha\beta}\, g^{\alpha l}\, g^{\beta m}.$$

By the general property of the principal composition the following equation holds:

$$G_{iklm} = G^{\lambda\mu\nu\rho}\, g_{i\lambda}\, g_{k\mu}\, g_{l\nu}\, g_{m\rho}$$

Therefore

$$G_{ik}^{lm} = G^{\lambda\mu\nu\rho}\, g_{ik}\, g_{k\mu}\, g_{\alpha\nu}\, g_{\beta\rho}\, g^{\alpha l}\, g^{pm} = G^{\lambda\mu lm}\, g_{i\lambda}\, g_{k\mu} = \frac{1}{\sqrt{g}}\, \delta_{\lambda\mu lm}\, g_{i\lambda}\, g_{k\mu}.$$

The quantities δ_{iklm} do not form a tensor; we will prove that it is easy from them to form the tensor $\mathbf{L}_{ik}^{lm}$ in the following way:

$$L_{ik}^{lm} = \delta_{\alpha\beta ik}\, \delta_{\alpha\beta lm},$$

In fact, from the previous we have:

$$L_{ik}^{lm} = \sqrt{g}\, \delta_{\alpha\beta ik} \frac{1}{\sqrt{g}}\, \delta_{\alpha\beta lm} = G_{\alpha\beta ik}\, G^{\alpha\beta lm}.$$

In such a way $\mathbf{L}_{ik}^{lm}$ is formed by the composition of two tensors and is therefore also a tensor.

4   Tensor Analyses

1. Turning to the study of the rules of tensor analysis, we face a significant difficulty and the attempts to overcome it led to the development of the modern tensor calculus. We could think, that, forming the usual partial derivatives with respect to the variables x_k of different components of some tensor, we will again obtain a tensor, but of a higher rank. In fact, it is not so – forming the usual derivatives of a tensor does not give us quantities, possessing a tensor character; It is necessary, as it was for the first time done by Richi, to modify the notion of a derivative, in order that in the operation of differentiating tensors the tensor character of the resulting quantities is not lost.

Assume that for every element of the manifold M_n we know the contragradient vector $\mathbf{a}_i$, we will say that we have *a field of contragradient vectors* $\mathbf{a}_i$. It is obvious that $\mathbf{a}_i$ will be functions of $x_1, x_2, \cdots x_n$; we will not talk much about the analogous character of these functions, assuming that this functions are continuous and have as many continuous derivatives as we need in the study of relevant questions.

Let us form the quantities $a_{i,k}$ by the equations:

$$a_{i,k} = \frac{\partial a_i}{\partial x_k}$$

We will show, that the quantities a_{ik} do not have a tensor character; using transformations of the variables, we find the quantities $a_{i,k}$:

$$\overline{a}_{i,k} = \frac{\partial \overline{a}_i}{\partial \overline{x}_k},$$

but, by the definition of the contragradient vector $\mathbf{a}_i$ we will have:

$$\overline{a}_i = a_x \frac{\partial x_k}{\partial \overline{x}_i},$$

108

from where

$$\frac{\partial \overline{a}_i}{\partial \overline{x}_k} = \frac{\partial(a_\alpha)}{\partial \overline{x}_k}\frac{\partial x_\alpha}{\partial \overline{x}_i} + a_\alpha \frac{\partial^2 x_\alpha}{\partial \overline{x}_i \partial \overline{x}_k} = \frac{\partial a_\alpha}{\partial x_\beta}\frac{\partial x_\alpha}{\partial \overline{x}_i}\frac{\partial x_\beta}{\partial \overline{x}_k} + a_\alpha \frac{\partial^2 x_\alpha}{\partial \overline{x}_i \partial \overline{x}_k}$$

or

$$\overline{a}_{ik} = a_{\alpha,\beta}\frac{\partial x_\alpha}{\partial \overline{x}_i}\frac{\partial x_\beta}{\partial \overline{x}_k} + a_\alpha \frac{\partial^2 x_\alpha}{\partial \overline{x}_i \partial \overline{x}_k},$$

It is obvious that the second sum in the above equation does not disappear, which shows that the quantities a_{ik} do not have tensorial character.

A closer examination of the derivatives a_{ik} shows that it is possible to form a combination of $a_{i,k}$ and of the vector $\mathbf{a}_l$ itself which would have tensorial character. In order to form such a combination it is necessary to use the special quantities , depending on three indices and *not having* tensorial character. Further we will see how by using the fundamental tensor we can form those quantities; now we will examine their properties in the general form.

We will call tensorial parameters the set of quantities $\Gamma^i_{\lambda\mu}$ $(i, \lambda, \mu = 1, 2, \cdots n)$, *defined for each of the manifolds* M_n *and transforming into the quantities* $\overline{\Gamma}^i_{\lambda\mu}$ *in another manifold* $\overline{M}_n$ *by the formulas:*

$$\overline{\Gamma}^i_{\lambda\mu} = \Gamma^\sigma_{\alpha\beta}\frac{\partial x_\alpha}{\partial \overline{x}_\lambda}\frac{\partial x_\beta}{\partial \overline{x}_\beta}\frac{\partial \overline{x}_i}{\partial x_\sigma} + \frac{\partial^2 x_\alpha}{\partial \overline{x}_\lambda \partial \overline{x}_\mu}\frac{\partial \overline{x}_i}{\partial x_\sigma}. \tag{27}$$

Multiplying the two sides of the formula (27) by $\frac{\partial x_\nu}{\partial \overline{x}_i}$ and summing over i we will express $\frac{\partial^2 x_\nu}{\partial \overline{x}_\lambda \partial \overline{x}_\mu}$ through the tensorial parameters in M_n and $\overline{M}_n$; in fact:

$$\frac{\partial^2 x_\sigma}{\partial \overline{x}_\lambda \partial \overline{x}_\mu}\frac{\partial x_i}{\partial x_\sigma}\frac{\partial x_\nu}{\partial \overline{x}_i} = \frac{\partial^2 x_\sigma}{\partial \overline{x}_\lambda \partial \overline{x}_\mu}\delta^\nu_\sigma = \frac{\partial^2 x_\nu}{\partial \overline{x}_\lambda \partial \overline{x}_\mu}.$$

Hence, from formula (27) we get:

$$\frac{\partial^2 x_\nu}{\partial \overline{x}_\lambda \partial \overline{x}_\mu} = \overline{\Gamma}^i_{\lambda\mu}\frac{\partial x_\nu}{\partial \overline{x}_i} - \Gamma^\nu_{\alpha\beta}\frac{\partial x_\alpha}{\partial \overline{x}_\lambda}\frac{\partial x_\beta}{\partial \overline{x}_\mu}. \tag{28}$$

In the same way, replacing x_i by $\overline{x}_i$ and vice-versa, we find the following equation:

$$\frac{\partial^2 \overline{x}_\nu}{\partial x_\lambda \partial x_\mu} = \Gamma^i_{\lambda\mu}\frac{\partial \overline{x}_\nu}{\partial x_i} - \overline{\Gamma}^\nu_{\alpha\beta}\frac{\partial \overline{x}_\alpha}{\partial x_\lambda}\frac{\partial \overline{x}_\beta}{\partial x_\mu}. \tag{29}$$

We will show, how with the help of tensorial parameters, it is possible to form expressions from the contragradient vector $\mathbf{a}_i$ and its derivatives which have tensorial character. We will rewrite the formula for $\overline{a}_{ik}$ in the following way, replacing in it i, k by λ and $\mu.$, and the arameter α in the second term by ν:

$$\overline{a}_{\lambda,\mu} = a_{\alpha,\beta}\,\frac{\partial x_\alpha}{\partial \overline{x}_\lambda}\,\frac{\partial x_\beta}{\partial \overline{x}_\mu} + a_\nu\,\frac{\partial^2 x_\nu}{\partial \overline{x}_\lambda \partial \overline{x}_\mu}.$$

For $\frac{\partial^2 x_\nu}{\partial \overline{x}_\lambda \partial \overline{x}_\mu}$ from formula (28) we will have:

$$\overline{a}_{\lambda,\mu} = a_{\alpha,\beta}\,\frac{\partial x_\alpha}{\partial \overline{x}_\lambda}\,\frac{\partial x_\beta}{\partial \overline{x}_\mu} + a_\nu\,\frac{\partial x_\nu}{\partial x_i}\,\overline{\Gamma}^i_{\lambda\mu} - a_\nu\,\Gamma^\nu_{\alpha\beta}\,\frac{\partial x_\alpha}{\partial \overline{x}_\lambda}\,\frac{\partial x_\beta}{\partial \overline{x}_\mu},$$

but, by the property of the contragradient vector $\mathbf{a}_\nu$, we will have:

$$\overline{a}_i = a_\nu\,\frac{\partial x_\nu}{\partial \overline{x}_i},$$

from where

$$\overline{a}_{\lambda\mu} - \overline{a}_i\,\overline{\Gamma}^i_{\lambda\mu} = (a_{\alpha\beta} - a_\nu\,\Gamma^\nu_{\alpha\beta})\,\frac{\partial x_\alpha}{\partial \overline{x}_\lambda}\,\frac{\partial x_\beta}{\partial \overline{x}_\mu}.$$

The last equation shows that the expression

$$a'_{ik} = a_{i,k} - a_\sigma\,\Gamma^\sigma_{ik}$$

is a contragradient tensor of the second rank, because

$$\overline{a}'_{ik} = a'_{\alpha\beta}\,\frac{\partial x_\alpha}{\partial \overline{x}_i}\,\frac{\partial x_\beta}{\partial \overline{x}_k}.$$

In such way, with the help of tensorial parameters, it is not difficult to form differential operations over tensors possessing tensorial character. Before considering this question in the general case, we will point out some properties of tensorial parameters, properties, which we will need later, whose derivation is very simple but needs a lot of calculations.

The following question arises naturally: if for M_n a system of parameters $\Gamma^i_{\lambda\mu}$ is given and if another system of parameters $\overline{\Gamma}^i_{\lambda\mu}$ is given, then is it possible always to find such a transformation of variables, where, by the given formulas of tensorial parameters, we can go from the parameters $\Gamma^i_{\lambda\mu}$ to the parameters $\overline{\Gamma}^i_{\lambda\mu}$? It turns out, that for the possibility of such a transition it is necessary that some conditions should be satisfied, namely, the two expressions should possess tensorial character – one should be a mixed tensor of the third rank – the other – a mixed tensor of the fourth rank.

The first equation is defined by the equation:

$$\gamma^i_{\lambda\mu} = \Gamma^i_{\lambda\mu} - \Gamma^i_{\mu\lambda}. \tag{30}$$

The equation has no name because in most applications the tensorial parameters are symmetrical with respect to low indices and therefore $\gamma^i_{\lambda\mu}$ simply becomes zero. We will call this equation *symmetral.*

The second equation is of a more complex composition, depending on four indices and defined by the formula:

$$F^i_{k\lambda\mu} = \frac{\partial \Gamma^i_{k\lambda}}{\partial x_\mu} - \frac{\partial \Gamma^i_{k\mu}}{\partial x_\lambda} + \Gamma^\sigma_{k\lambda}\,\Gamma^i_{\sigma\mu} - \Gamma^\sigma_{k\mu}\Gamma^i_{\sigma\lambda}. \tag{31}$$

The set of these equations,which have special importance in tensor calculus as well as in its applications, has the name *curvature of tensorial parameters.* Since it should have tensorial character its name should be *tensor of the curvature tensorial parameters.*

Theorem 7 *In order that, the system of n^3 quantities $\Gamma^i_{\lambda\mu}$ can with the help of transformations of variables change into the system of n^3 quantities $\overline{\Gamma}^i_{\lambda\mu}$ by using the formulas of transformations, corresponding to the transformation of tensorial parameters, it is necessary and sufficient that:*

1) the quantities $\gamma^i_{\lambda\mu}$ possess for the given transformation of variables a tensorial character (cogradient for the upper and contragradient for low indices) and

2) the curvature $\Gamma^i_{k\lambda\mu}$ possesses for the mentioned transformation of variables a tensorial character (cogradient for upper and contragradient for low indices).

Before proving this theorem we notice that for tensorial parameters the tensor nature of $\gamma^i_{\lambda\mu}$ and the curvature of tensorial parameters follows directly from theorem 7.

The proof of this theorem actually shows that its conditions follow from equations (27) or from (29), which are equivalent to (27), and that, vice-versa, as long as the conditions of theorem 7 are satisfied then it is always possible to find $\overline{x}_\nu$ as functions of $x_1, x_2, \cdots x_n$ satisfying equations (29).

Let us prove firstly the necessary condition of the theorem.

Exchanging the places of λ and μ in equations (29) and subtracting the obtained equations we find:

$$\gamma^i_{\lambda\mu} \frac{\partial \overline{x}_\nu}{\partial x_i} - \overline{\gamma}^\nu_{\alpha\beta} \frac{\partial \overline{x}_\alpha}{\partial x_\lambda} \frac{\partial \overline{x}_\beta}{\partial x_\mu} = 0,$$

and multiplying by $\frac{\partial x_j}{\partial \overline{x}_\nu}$ and summing up over ν we find:

$$\gamma^j_{\lambda\mu} = \overline{\gamma}^\nu_{\alpha\beta} \frac{\partial \overline{x}_\alpha}{\partial x_\lambda} \frac{\partial \overline{x}_\beta}{\partial x_\mu} \frac{\partial x_j}{\partial \overline{x}_\nu}.$$

This equation shows the tensorial character of $\gamma^j_{\lambda\mu}$.

In order to get the conditions about the curvature, we rewrite equations (29) in the following way:

$$\frac{\partial^2 \overline{x}_i}{\partial x_\lambda \partial x_\mu} = \Gamma^\beta_{\lambda\mu} \frac{\overline{x}_i}{\partial x_\beta} - \overline{\Gamma}^i_{\alpha\beta} \frac{\partial \overline{x}_\alpha}{\partial x_\lambda} \frac{\partial \overline{x}_\beta}{\partial x_\mu};$$

$$\frac{\partial^2 \overline{x}_i}{\partial x_\lambda \partial x_\nu} = \Gamma^\beta_{\lambda\nu} \frac{\overline{x}_i}{\partial x_\beta} - \overline{\Gamma}^i_{\alpha\beta} \frac{\partial \overline{x}_\alpha}{\partial x_\lambda} \frac{\partial \overline{x}_\beta}{\partial x_\nu}.$$

Differentiating the first of the obtained equations with respect to x_ν, and the second with respect to x_μ, replacing everywhere the second derivatives by the first according to formulas (29) and subtracting the obtained equations, after some simplifications, we will have:

$$F^\beta_{\lambda\mu\nu} \frac{\partial \overline{x}_i}{\partial x_\beta} = \overline{F}^i_{\alpha\beta\gamma} \frac{\partial \overline{x}_\alpha}{\partial x_\lambda} \frac{\partial \overline{x}_\beta}{\partial x_\mu} \frac{\partial \overline{x}_\gamma}{\partial x_\nu}$$

or

$$F^k_{\lambda\mu\nu} = \overline{F}^i_{\alpha\beta\gamma} \frac{\partial \overline{x}_\alpha}{\partial x_\lambda} \frac{\overline{x}_\beta}{\partial x_\mu} \frac{\partial \overline{x}_\gamma}{\partial x_\nu} \frac{\partial x_k}{\partial \overline{x}_i},$$

which shows the tensorial character of the curvature.

Now we will show the sufficient condition for our theorem. The fact that the conditions of the theorem are satisfied makes it possible to obtain from equation (29) $\frac{\partial^2 \overline{x}_i}{\partial x_\lambda \partial x_\mu}$ and $\frac{\partial^2 \overline{x}_i}{\partial x_\lambda \partial x_\mu \partial x_\nu}$, where these quantities turn out to be equal regardless of the order in which we differentiate them. It is not difficult to see that when these conditions hold the system of equations (29) will be integrable system, i.e., by a given number of differentiation it is always possible to determine from equations (29) any differentiation of any order of $\overline{x}_i$; so the value of this derivative will be the same regardless of how we changed the order of differentiation. Once this point is clarified, then, according to the general theory of differential equations, we can assert the existence of solutions of the system (29) (the so called solution in the sense of Cauchy).

Let us prove, that the derivatives of the fourth order $\overline{x}_i$ with respect to x_λ, x_μ, x_ν, x_k obtained from equation (29) will not depend

112

on the order of differentiation. We will prove, for example, that, determining $\frac{\partial^4 \overline{x}_i}{\partial x_\lambda \partial x_\mu \partial x_\nu \partial x_k}$ from equation (29), we obtain the same result as determining $\frac{\partial^4 \overline{x}_i}{\partial x_k \partial x_\mu \partial x_\nu \partial x_\lambda}$ from these equations; we have:

$$\frac{\partial^4 \overline{x}_i}{\partial x_\lambda \partial x_\mu \partial x_\nu \partial x_k} = \frac{\partial}{\partial x_k} \left(\frac{\partial^3 \overline{x}_i}{\partial x_\lambda \partial x_\mu \partial x_\nu} \right),$$

$$\frac{\partial^4 \overline{x}_i}{\partial x_k \partial x_\mu \partial x_\nu \partial x_\lambda} = \frac{\partial}{\partial x_\lambda} \left(\frac{\partial^3 \overline{x}_i}{\partial x_k \partial x_\mu \partial x_\nu} \right),$$

but we just proved that

$$\frac{\partial^3 \overline{x}_i}{\partial x_\lambda \partial x_\mu \partial x_\nu} = \frac{\partial^3 \overline{x}_i}{\partial x_\mu \partial x_\nu \partial x_\lambda}, \quad \frac{\partial^3 \overline{x}_i}{\partial x_k \partial x_\mu \partial x_\nu} = \frac{\partial^3 \overline{x}_i}{\partial x_\mu \partial x_\nu \partial x_k},$$

therefore:

$$\frac{\partial^4 \overline{x}_i}{\partial x_\lambda \partial x_\mu \partial x_\nu \partial x_k} = \frac{\partial^2}{\partial x_k \partial x_\lambda} \left(\frac{\partial^2 \overline{x}_i}{\partial x_\mu \partial x_\nu} \right),$$

$$\frac{\partial^4 \overline{x}_i}{\partial x_k \partial x_\mu \partial x_\nu \partial x_\lambda} = \frac{\partial^2}{\partial x_\lambda \partial x_k} \left(\frac{\partial^2 \overline{x}_i}{\partial x_\mu \partial x_\nu} \right).$$

But the right sides of these equations are obviously identical, therefore the left sides are also equal, which is what we wanted to proof.

We will also prove, that derivatives of any order $\overline{x}_i$ will be the same, regardless of the order of differentiation of equations (29). Once this is proved, then, according to the above remark, the reverse theorem will hold.

The first condition of theorem 7, which is not difficult to see, is eliminated as soon as the tensorial parameters are symmetric with respect to the low indices. We call such tensorial parameters *symmetric*. Obviously, the necessary and sufficient condition for bringing all symmetric tensorial parameters to zero, will be the vanishing of the curvatures of these tensorial parameters. In fact if all tensorial parameters for any $\overline{M}_n$ are equal to zero, then obviously $F^i_{k\lambda\mu}$ for this $\overline{M}_n$ will be equal to zero as well. As far as it is a tensor then it will be equal to zero for any other M_n. The symmetric parameters whose curvature is zero will be called *zero parameters*.

Direct calculations show that the curvature of tensorial parameters is a tensor which is antisymmetric in the last two low indices, because we have the equation:

$$F^i_{k\lambda\mu} = -F^i_{k\mu\lambda}. \tag{32}$$

In the same way for symmetric tensorial parameters, the tensor of the curvature has the property *cyclical symmetry with respect to three low indices*; in fact for symmetric parameters we have:

$$F^i_{k\lambda\nu} + F^i_{\lambda\mu k} + F^i_{\mu k\lambda} = 0. \tag{33}$$

With the help of the principal and the unit composition of the curvature of tensorial parameters we can form two tensors, which will be needed later, namely: the *Riemann tensor* $\mathbf{F}_{iklm}$ and *the contracted Riemann tensor* $\mathbf{F}_{ik}$ of the tensorial parameters. The Riemann tensor is obtained with the help of the principal composition:

$$F_{iklm} = F^\alpha_{klm} g_{\alpha i} \tag{34}$$

and vice-verse

$$F^i_{k\lambda\mu} = g^{i\sigma} F_{\sigma k\lambda\mu}. \tag{35}$$

In the same way the contraction of the Riemann tensor is formed with the help of the unit composition:

$$F_{ik} = F^\alpha_{i\alpha k}. \tag{36}$$

It is not difficult to go from the contracted Riemann tensor to the scalar F called *scalar curvature* of the tensorial parameters:

$$F = g^{ik} F_{ik}. \tag{37}$$

2. Let us now consider the definition of the so called *tensorial derivatives* of different tensors. In order to avoid unnecessary calculations we, using tensorial parameters, introduce a special system of co- and contragradient vectors with whose help we will create form tensorial derivatives.

Choosing for x_i arbitrary functions of some variable t, we can form completely arbitrarily cogradient vector $\mathbf{v}^i$:

$$v^i = \frac{dx_i}{dt}.$$

As only x_i will be given functions of t, then the tensorial parameters, which are functions of $x_1, x_2 \cdots x_n$, will be functions of t. We will form the following system of n linear differential equations with n unknown functions t:

$$\xi^1, \xi^2, \cdots \xi^n;$$

$$\frac{d\xi^i}{dt} = -\Gamma^i_{rs} \frac{dx_s}{dt} \xi^r = -\Gamma^i_{rs} v^s \xi^r, \quad i = 1, 2, \cdots n. \tag{38}$$

114

According to the general theory of linear differential equations, the system (38) has a solution, transforming at $t = t_0$ (the initial value of t) into given in advance constant quantities $^0\xi^i$; we clarify that such a solution can be regarded as a cogradient vector, as long as $^0\xi^i$ is a a cogradient vector. For M_n, ξ^i are defined as solutions of the system of equations and initial conditions:

$$\frac{d\xi^i}{dt} = -\Gamma^i_{rs} \frac{dx_s}{dt}\, \xi^r = -\Gamma^i_{rs}\, v^s\, \xi^r, \quad (i = 1, 2, \cdots n), \qquad (*)$$

$$\xi^i =^0 \xi^i, \text{ when } t = t_0.$$

In exactly the same way we have for $\overline{M}_n$:

$$\frac{d\bar{\xi}^i}{dt} = -\overline{\Gamma}^i_{rs} \frac{d\bar{x}_s}{dt}\, \bar{\xi}^r = -\overline{\Gamma}^i_{rs}\, \bar{v}^s\, \bar{\xi}^r, \qquad (**)$$

$$\bar{\xi}^i =^0 \bar{\xi}^i, \text{ when } t = t_0.$$

It is necessary, knowing that $^0\xi^i$ is a cogradient vector, to prove the vector nature of $\bar{\xi}^i$. We replace i by α in equation $(*)$ and in equation $(**)$ r by λ, s by μ, multiply both sides of the equation by $\frac{\partial \bar{x}_i}{\partial x_\alpha}$, sum up over α from I to n and subtract the result from equation $(**)$; then we obtain:

$$\frac{d\xi^i}{dt} - \frac{d\bar{x}_i}{dx_\alpha}\frac{d\xi^\alpha}{dt} = -\overline{\Gamma}^i_{\lambda\mu}\, \bar{v}^\mu\, \bar{\xi}^\lambda + \Gamma^\alpha_{rs}\frac{\partial \bar{x}_i}{\partial x_\alpha}\, v^s\, \xi^r,$$

but by formula (29) we have:

$$\Gamma^\alpha_{rs}\frac{\partial x_i}{\partial x_\alpha} = \frac{\partial^2 x_i}{\partial x_r \partial x_s} + \overline{\Gamma}^i_{\alpha\beta}\frac{\partial \bar{x}_\alpha}{\partial x_r}\frac{\partial \bar{x}_\beta}{\partial x_s}.$$

Therefore the previous equation will be rewritten in this way:

$$\frac{d\xi^i}{dt} - \frac{\partial \bar{x}_i}{\partial x_\alpha}\frac{d\xi^\alpha}{dt} - \frac{\partial^2 \bar{x}_i}{\partial x_r \partial x_s}\, v^s\, \xi^r = -\overline{\Gamma}^i_{\lambda\mu}\, \bar{v}^\mu\, \bar{\xi}^\lambda + \overline{\Gamma}^i_{\alpha\beta}\frac{\partial \bar{x}_\alpha}{\partial x_r}\frac{\partial \bar{x}_\beta}{\partial x_s}\, v^s\, \xi^r,$$

As $\bar{v}^\beta = \frac{\partial \bar{x}_\beta}{\partial x_s}\, v^2$ and replacing β on the right by μ and α by λ we find:

$$\frac{\xi^i}{dt} - \frac{\bar{x}_i}{\partial x_\alpha}\frac{d\xi^\alpha}{dr} - \frac{\partial^2 \bar{x}_i}{\partial x_r \partial x_s}\, v^s \xi^r = -\overline{\Gamma}^i_{\lambda\mu}\, \bar{v}^\mu = -\overline{\Gamma}^i_{\lambda\mu}\, \bar{v}^\mu\left(\bar{xi}^\lambda - \frac{\partial \bar{x}_\lambda}{\partial x_r}\, \xi^r\right).$$

As $v^s = \frac{dx_s}{dt}$ and replacing in the left-hand side of the equation r by α we obtain:

$$\frac{d\xi^i}{dt} = -\overline{\Gamma}^i_{\lambda\mu}\, \bar{v}^\mu\, \xi^\lambda, \qquad (***)$$

where:

$$\xi^i = \bar{\xi}^i - \frac{\partial \bar{x}_i}{\partial x_\alpha} \xi^\alpha.$$

As $^0\xi^i$ is a cogradient vector and therefore when $t = t_0$ ξ^i becomes zero and as the functions ξ^i satisfy the system of linear equations $(***)$, having a single solution with the given initial conditions, then ξ^i is equal to zero for all t, i.e.:

$$\bar{\xi}^i = \xi^\alpha \frac{\partial \bar{x}_i}{\partial x_\alpha},$$

which proves the vector nature of the introduced expression for ξ^i.

In such a way it is always possible to find cogradient vector ξ^i, becoming, when $t = t_0$, in any in advance given vector $^0\xi^i$ and satisfying the system of equations (38).

In the same way, *as long as $G^i_{\lambda\mu}$ are tensorial parameters, then it is always possible to find a contragradient vector η_i, becoming, when $t = t_0$, in any in advance given vector $^0\eta_i$ and satisfying the following system of equations:*

$$\frac{d\eta_i}{dt} = G^r_{is} \frac{dx_s}{dt} \eta_r = G^r_{is} v^s \eta_r. \tag{39}$$

The proof of this second assertion is completely analogous to the first one and for this reason we will skip it. In the next chapter we will see, when studying parallel transport in curved spaces, what will be the role of the vectors introduced by equations (38) and (39).

With the help of these vectors we can form an extremely useful method for obtaining tensor derivatives. Let us form the already mentioned above tensor derivative of a contragradient vector $\mathbf{a}_i$. For that reason we combine it with some vector ξ^i, satisfying the system (38) and as a result of the composition we get the scalar Φ:

$$\Phi = a_i \xi^i = \bar{a}_i \bar{\xi}^i.$$

Differentiating this scalar with respect to t we get:

$$\frac{d}{dt}(a_i \xi^i) = \frac{\bar{d}}{dt}(\bar{a}_i \xi^i).$$

Doing the differentiation, since

$$\frac{da_i}{dt} = \frac{\partial a_i}{\partial x_s} \frac{dx_s}{dt} = \bar{a}_{i,s} v^2,$$

116

$$\frac{d\overline{a}_i}{dt} = \frac{\partial \overline{a}_i}{\partial \overline{x}_s} \frac{d\overline{x}_s}{dt} = \overline{a}_{i,s}\, v^{\overline{s}}$$

and using equation (38) for $\frac{d\xi^i}{dt}$ and properly modified for $\frac{d\overline{\xi}^i}{dt}$ we will have:

$$a_{i,s}\, v^s\, \xi^i - a_i\, \Gamma^i_{rs}\, v^s\, \xi^r = \overline{a}_{i,s}\, \overline{v}^s\, \overline{\xi}^i - \overline{a}_i\, \overline{\Gamma}^i_{r,s}\, v^{\overline{s}}\, \overline{\xi}^r,$$

or

$$\left(a_{i,k} - a_\sigma\, \Gamma^\sigma_{i,k}\right) \xi^i\, v^k = \left(\overline{a}_{i,k} - \overline{a}_\sigma\, \overline{\Gamma}^\sigma_{ik}\right) \overline{\xi}^i\, \overline{v}^k,$$

i.e.,

$$a'_{ik}\, \xi^i\, v^k$$

is a scalar, and as v^k is an arbitrary vector, and ξ^i, due to the arbitrariness of $^0\xi^i$, is also arbitrary at $t = t_0$, then from here it follows that

$$a'_{ik} = a_{i,k} - a_\sigma\, \Gamma^\sigma_{ik}$$

is a contragradient tensor of the second rank.

We will denote this tensor by the sign $\mathbf{a}_{i\cdot k}$, so

$$\mathbf{a}_{i\cdot k} = a'_{ik} = a_{i,k} - a_\sigma\, \Gamma^\sigma_{ik}.$$

In such a way we obtain the derivatives, by forming from the given tensor and vectors, satisfying the equations (38) and (39), a scalar, then differentiating this scalar with respect to t and, with the help of equations (38) and (39), bring it in the form of some composition of the vector v^s, vectors, satisfying equations (38) and (39) and as an expression dependent on indices; the tensorial character of this expression will follow from the fact that the obtained composition is a scalar, and v^s and the vectors satisfying equations (38) and (39) will be arbitrary vectors.

Let us find in this way the derivative of the cogradient vector $\mathbf{a}^i$: let $a^i_{(k)} = \frac{\partial a^i}{\partial x_k}$.

We form the scalar:

$$\Phi = a^i\, \eta_i,$$

and differentiate it with respect to t:

$$\frac{d\Phi}{dt} = \frac{\partial a^i}{\partial x_s} \frac{dx_s}{dt}\, \eta_i + a^i\, \frac{d\eta_i}{dt} = a^i_{(s)}\, v^s\, \eta_i + a^i\, G^r_{is}\, v^s\, \eta_r = \left(a^i_{(s)} + a^\sigma\, G^i_{\sigma s}\right) v^s\, \eta_i.$$

Introducing the notation:

$$\mathbf{a}^i_k = a^i_{(k)} + a^\sigma\, G^i_{\sigma k},$$

we find that $\overset{\xi\,i}{a_k}$ is a tensor of the second rank, which is the tensor derivative of the vector $\mathbf{a}^i$.

Let us denote this tensor by

$$\mathbf{a}^i_{.k},$$

i.e.

$$a^i_{.k} = \overset{\xi\,i}{a_k} = a^i_{(k)} + a^\sigma\, G^i_{\sigma k}.$$

Now we find the tensor derivative of the tensor $\mathbf{a}_{ik}$.

Let ξ^i and ζ^i be two vectors satisfying equations (38). We form the scalar:

$$\Phi = a_{ik}\,\xi^i \zeta^k,$$

and differentiate it with respect to t

$$\frac{d\Phi}{dt} = a_{ik\cdot s}\, v^s\, \xi^i\, \zeta^k - a_{ik}\,\Gamma^i_{rs}\, v^s\, \xi^r\, \zeta^k - a_{ik}\,\xi^i\,\Gamma^k_{rs}\, v^s\, \zeta^r =$$

$$= v^s\, \xi^i\, \zeta^k\, \left(a_{ik\cdot s} - a_{\sigma k}\,\Gamma^\sigma_{is} - a_{i\sigma}\,\Gamma^\sigma_{ks}\right).$$

So the tensor:

$$a_{ik\cdot l} = a_{ik,l} - a_{\sigma k}\,\Gamma^\sigma_{il} - a^\sigma_{i\sigma}\,\Gamma^\sigma_{kl},$$

will be the tensor derivative of the tensor $\mathbf{a}_{ik}$.

In order to define the tensor derivative in the general case we will introduce several notations. We will denote the ordinary derivative of the components of the tensor $T^{i_1,i_2\cdots i_s}_{j_1,j_2\cdots j_s}$ with respect to x_l by the symbol $T^{i_1 i_2\cdots i_s}_{j_1 j_2\cdots j_s,l}$

$$\frac{\partial T^{i_1 i_2\cdots i_s}_{j_1 j_2\cdots j_s}}{\partial x_l} = T^{i_1 i_2\cdots i_s}_{j_1 j_2\cdots j_s,l}$$

In the case of absence of low indices we will write a comma before the index, indicating the letter on which the derivative is taken.

$$\frac{\partial T^{i_1 i_2\cdots i_r}}{\partial x_l} = T^{i_1 i_2\cdots r_i}_{,l}$$

The tensor derivative of the tensor $T^{i_1 i_2\cdots i_r}_{j_1 j_2\cdots j_s}$ with respect to x_l, i.e., such a combination of its derivatives with respect to x_l and its components and tensorial parameters, which itself is a tensor will be denoted by the notation:

$$\mathbf{T}^{i_1 i_2\cdots i+r}_{j_1 j_2\cdots j_s\cdot l}$$

118

In order to form this derivative we choose s cogradient ${}^j\xi^i$ ($j = 1, 2 \cdots s$) and r contragradient vectors ${}^j\eta_i$ ($j = 1, 2 \cdots r$), satisfying equations (38) and (39), we form the scalar:

$$\Phi = T^{i_1 i_2 \cdots i_r}_{j_1 j_2 \cdots j_s} {}^1\xi^{j_1} \, {}^2\xi^{j_2} \cdots {}^s\xi^{j_s} \, {}^1\eta_{i_1} \, {}^2\eta_{i_2} \cdots {}^r\eta_{i_r}.$$

Differentiating this scalar with respect to t and using equations (38) and (39) we find the following expression for the tensor derivative:

$$T^{i_1 i_2 \cdots i_r}_{j_1 j_2 \cdots j_s \cdot l} = T^{i_1 i_2 \cdots i_r}_{j_1 j_2 \cdots j_s, l} - T^{i_1 i_2 \cdots i_r}_{\sigma_1 j_2 \cdots j_s} \, \Gamma^{\sigma_1}_{j_1 l} - T^{i_1 i_2 \cdots i_r}_{j_1 \sigma_2 \cdots j_s} \, \Gamma^{\sigma_2}_{j_2 l} - \cdots$$

$$-T^{i_1 i_2 \cdots i_r}_{j_1 j_2 \cdots \sigma_s} \, \Gamma^{\sigma_s}_{j_s l} + T^{s_1 i_2 \cdots i_r}_{j_1 j_2 \cdots j_s} \, G^{i_1}_{s_1 l} + T^{i_1 s_2 \cdots i_r}_{j_1 j_2 \cdots j_s} \, G^{i_2}_{s_2 l} + \cdots$$

$$+ T^{i_1 i_2 \cdots s_r}_{j_1 j_2 \cdots j_s} \, G^{i_r}_{s_r l}. \tag{40}$$

Obviously, from the tensor $\mathbf{T}^{i_1 i_2 \cdots i_r}_{j_1 j_2 \cdots j_s \cdot l}$ it is also possible to form a tensor derivative with respect to x_k which we will denote by $\mathbf{T}^{i_1 i_2 \cdots i_r}_{j_1 j_2 \cdots j_s \cdot lk}$. Generally speaking, following this procedure further, it is possible to form a tensor derivative of m-th rank from the tensor $\mathbf{T}^{i_1 i_2 \cdots i_r}_{j_1 j_2 \cdots j_s}$ with respect to $x_{l_1}, x_{l_2}, \cdots x_{l_m}$ which will be denoted by the following notation: $\mathbf{T}^{i_1 i_2 \cdots i_r}_{j_1 j_2 \cdots j_s \cdot l_1 l_2 l_m}$.

We will not derive general formulas for obtaining this derivative of m-th rank, but will consider only some special cases.

We will find the derivative of a contragradient vector of the second rank $\mathbf{a}_i$ with respect to x_k and x_l. We have:

$$a_{i \cdot k} = a_{i \cdot k} - a_\sigma \, \Gamma^\sigma_{ik},$$

$$a_{i \cdot kl} = a_{l \cdot k, l} - a_{\sigma \cdot k} \, \Gamma^\sigma_{kl},$$

but

$$a_{i \cdot k, l} = a_{i, kl} - a_{\sigma, l} \, \Gamma^\sigma_{ik} - a_\sigma \, \frac{\partial \Gamma^\sigma_{ik}}{\partial x_l},$$

$$a_{\sigma \cdot k} = a_{\sigma, k} - a_\rho \, \Gamma^\rho_{\sigma k}, \quad a_{i \cdot \sigma} = a_{i, \sigma} - a_\rho \, \Gamma^\rho_{i\sigma}.$$

Therefore

$$a_{i \cdot kl} = a_{i, kl} - a_{\sigma, l} \, \Gamma^\sigma_{ik} - a_{\sigma, k} \, \Gamma^\sigma_{il} - a_{i, \sigma} \, \Gamma^\sigma_{kl} - a_\sigma \, \frac{\partial \Gamma^\sigma_{ik}}{\partial x_l} +$$

$$+ a_\rho \, \Gamma^\sigma_{il} \, \Gamma^\rho_{\sigma k} + a_\rho \, \Gamma^\sigma_{kl} \, \Gamma^\rho_{i\sigma}.$$

Changing the summation variable σ by ρ in the third term from the end, we find:

$$a_{i \cdot kl} = a_{i, kl} - a_{\sigma, k} \, \Gamma^\sigma_{il} - a_{i\sigma} \, \Gamma^\sigma_{kl} - a_{\sigma, l} \, \Gamma^\sigma_{ik}$$

$$-a_\rho \left(\frac{\partial \Gamma^\rho_{ik}}{\partial x_l} - \Gamma^\sigma_{il}\,\Gamma^\rho_{\sigma k} - \Gamma^\sigma_{kl}\,\Gamma^\rho_{i\sigma} \right).$$

By exchanging k and l we get:

$$a_{i\cdot lk} = a_{i,lk} - a_{\sigma,l}\,\Gamma^\sigma_{ik} - a_{i,\sigma}\,\Gamma^\sigma_{lk} - a_{\sigma,k}\,\Gamma^\sigma_{il} -$$

$$-a_\rho \left(\frac{\partial \Gamma^\rho_{il}}{\partial x_k} - \Gamma^\sigma_{ik}\,\Gamma^\rho_{\sigma l} - \Gamma^\sigma_{lk}\,\Gamma^\rho_{i\sigma} \right).$$

Subtracting from the formula, expressing $a_{i\cdot kl}$, the formula which gives $a_{i\cdot lk}$ we find:

$$a_{i\cdot kl} - a_{i\cdot lk} = -a_{i,\sigma}\,\gamma^\sigma_{kl} - a_\rho\,F^\rho_{ikl} + a_\rho\,\Gamma^\rho_{i\sigma}\,\gamma^\sigma_{kl} =$$

$$= -\left(a_{i,\sigma} - a_\rho\,\Gamma^\rho_{i\sigma} \right)\gamma^\sigma_{kl} - a_\rho\,F^\rho_{ikl},$$

from where:

$$a_\rho\,F^\rho_{ikl} = a_{i\cdot kl} - a_{i\cdot kl} + \gamma^\sigma_{kl}\,a_{i\cdot\sigma}.$$

The right-hand side of this equation is obviously a contragradient tensor of third rank for any vector $\mathbf{a}_\rho$, hence (see §1 of this chapter) $\mathbf{F}^\rho_{ikl}$ will be a mixed tensor of the fourth rank.

For symmetrical tensorial parameters we will have the relation:

$$a_{i\cdot ik} - a_{i\cdot kl} = a_\rho\,F^\rho_{ikl}.$$

3. While deriving the tensor derivatives we used special quantities, the so-called tensorial parameters; in the previous section we did not give even one example of forming these tensorial parameters and as a result the obtained tensor derivatives turned out so far to be only expressions with different symbols. A closer look at the fundamental tensor makes it possible to form a special type of tensorial parameters and also to find a general procedure for forming tensorial parameters. The introduced symbols will give at the same time, a significant for the geometry of curved spaces possibility to study the conditions for the transition from the fundamental tensor $\mathbf{g}_{ik}$, in one of the manifolds to the fundamental tensor $\mathbf{g}_{ik}$ in another manifolds.

Generally speaking, the components of the fundamental tensor g_{ik} will be functions of $x_1, x_2 \ldots x_n$; we assume that these functions are continuos and having continuous second partial derivatives with respect to the variables x_i. From the first derivatives g_{ik} we form an expression, called *square brackets of Christoffel* or *Christoffel brackets of the first kind*:

$$\begin{bmatrix} i\,k \\ l \end{bmatrix} = \frac{1}{2}\left(\frac{\partial g_{il}}{\partial x_k} + \frac{\partial g_{kl}}{\partial x_i} - \frac{\partial g_{ik}}{\partial x_l} \right) = \frac{1}{2}\left(g_{il,k} + g_{kl,i} - g_{ik,l} \right). \quad (41)$$

120

It turns out that with the help of this symbol it will be not be difficult to obtain the special form of the tensorial parameters.

We will clarify first of all how the square brackets of Christoffel are transformed under the transition from M_n to $\overline{M}_n$; denote the Christoffel brackets for $\overline{M}_n$ with a straight line over this symbol. We have:

$$\overline{\begin{bmatrix} i\,k \\ l \end{bmatrix}} = \frac{1}{2}\left(\frac{\partial \overline{g}_{il}}{\partial \overline{x}_k} + \frac{\partial \overline{g}_{kl}}{\partial \overline{x}_i} - \frac{\partial \overline{g}_{ik}}{\partial \overline{x}_l}\right) = \frac{1}{2}\left(\overline{g}_{il,k} + \overline{g}_{kl,i} - \overline{g}_{ik,l}\right).$$

Since:

$$\overline{g}_{il} = g_{\alpha\beta}\frac{\partial x_\alpha}{\partial \overline{x}_i}\frac{\partial x_\beta}{\partial \overline{x}_l},$$

then:

$$\overline{g}_{il,k} = g_{\alpha\beta,\gamma}\frac{\partial x_\alpha}{\partial \overline{x}_i}\frac{\partial x_\beta}{\partial \overline{x}_l}\frac{\partial x_\gamma}{\partial \overline{x}_k} + g_{\alpha\beta}\frac{\partial^2 x_\alpha}{\partial \overline{x}_i \partial \overline{x}_k}\frac{\partial x_\beta}{\partial \overline{x}_l} + g_{\alpha\beta}\frac{\partial x_\alpha}{\partial \overline{x}_i}\frac{\partial^2 x_\beta}{\partial \overline{x}_l \partial \overline{x}_k},$$

because

$$\frac{\partial g_{\alpha\beta}}{\partial \overline{x}_k} = \frac{\partial g_{\alpha\beta}}{\partial x_\gamma}\frac{\partial x_\gamma}{\partial \overline{x}_k} = g_{\alpha\beta,\gamma}\frac{\partial x_\gamma}{\partial \overline{x}_k}.$$

Changing respectively the index and forming $\overline{\begin{bmatrix} i\,k \\ l \end{bmatrix}}$ we find:

$$\overline{\begin{bmatrix} i\,k \\ l \end{bmatrix}} = \begin{bmatrix} i\,k \\ l \end{bmatrix}\frac{\partial x_\alpha}{\partial \overline{x}_i}\frac{\partial x_\beta}{\partial \overline{x}_k}\frac{\partial x_\gamma}{\partial \overline{x}_l} + g_{\alpha\beta}\frac{\partial x_\beta}{\partial \overline{x}_l}\frac{\partial^2 x_\alpha}{\partial \overline{x}_i \partial \overline{x}_k}. \tag{42}$$

Multiplying both parts of this equation by $\overline{g}^{sl}$ and summing over l from 1 to n we obtain the following equation:

$$\overline{g}^{sl}\begin{bmatrix} i\,k \\ l \end{bmatrix} = \begin{bmatrix} \alpha\,\beta \\ \gamma \end{bmatrix}\frac{\partial x_\alpha}{\partial \overline{x}_i}\frac{\partial x_\beta}{\partial \overline{x}_k}\overline{g}^{sl}\frac{\partial x_\gamma}{\partial \overline{x}_l} + g_{\alpha\beta}\frac{\partial^2 x_\alpha}{\partial \overline{x}_i \partial \overline{x}_k}\overline{g}^{sl}\frac{\partial x_\beta}{\partial \overline{x}_l}.$$

Since $\overline{g}^{ik}$ is a cogradient tensor we have the next equation:

$$\overline{g}^{ik} = g^{\alpha\beta}\frac{\partial \overline{x}_i}{\partial x_\alpha}\frac{\partial \overline{x}_k}{\partial x_\beta}.$$

Multiplying both parts of this equation by $\frac{\partial x_r}{\partial \overline{x}_k}$ and summing over k from 1 to n, we will have:

$$\overline{g}^{ik}\frac{\partial x_r}{\partial \overline{x}_k} = g^{\alpha\beta}\frac{\partial \overline{x}_i}{\partial x_\alpha}\frac{\partial \overline{x}_k}{\partial x_\beta}\frac{\partial x_r}{\partial \overline{x}_k} = g^{\alpha\beta}\frac{\partial \overline{x}_i}{\partial x_\alpha}\delta^r_\beta = g^{\alpha r}\frac{\partial \overline{x}_i}{\partial x_\alpha}.$$

In such a way

$$\overline{g}^{sl}\,\frac{\partial x_\gamma}{\partial \overline{x}_l} = g^{\sigma\gamma}\,\frac{\partial \overline{x}_s}{\partial x_\sigma}, \quad \overline{g}^{sl}\,\frac{\partial x_\beta}{\partial \overline{x}_l} = g^{\sigma\beta}\,\frac{\partial \overline{x}_s}{\partial x_\sigma}.$$

Inserting these expressions in the equation defining $\overline{g}^{sl}\,\overline{\begin{bmatrix} i\,k \\ l \end{bmatrix}}$, we find:

$$\overline{g}^{sl}\,\overline{\begin{bmatrix} i\,k \\ l \end{bmatrix}} = g^{\sigma\gamma}\,\begin{bmatrix} \alpha\,\beta \\ \gamma \end{bmatrix}\,\frac{\partial x_\alpha}{\partial \overline{x}_i}\,\frac{\partial x_\beta}{\partial \overline{x}_k}\,\frac{\partial \overline{x}_k}{\partial x_\sigma} + \frac{\partial^2 x_\alpha}{\partial \overline{x}_i \partial \overline{x}_k}\,\frac{\partial \overline{x}_s}{\partial x_\sigma}\,g_{\alpha\beta}\,g^{\sigma\beta} =$$

$$= g^{\sigma\gamma}\,\begin{bmatrix} \alpha\,\beta \\ \gamma \end{bmatrix}\,\frac{\partial x_\alpha}{\partial x\overline{x}_i}\,\frac{\partial x_\beta}{\partial \overline{x}_k}\,\frac{\partial \overline{x}_s}{\partial x_\sigma} + \frac{\partial^2 x_\alpha}{\partial \overline{x}_i \partial \overline{x}_k}\,\frac{\partial \overline{x}_s}{\partial x_\sigma}\,\delta_\alpha^\sigma =$$

$$= g^{\sigma\gamma}\,\begin{bmatrix} \alpha\,\beta \\ \gamma \end{bmatrix}\,\frac{\partial x_\alpha}{\partial \overline{x}_i}\,\frac{\partial x_\beta}{\partial \overline{x}_k}\,\frac{\partial \overline{x}_s}{\partial x_\sigma} + \frac{\partial^2 x_\sigma}{\partial \overline{x}_i \partial \overline{x}_k}\,\frac{\partial \overline{x}_s}{\partial x_\sigma}.$$

The obtained formula is similar to formula (27) with the help of which we defined the basic property of the tensorial parameters. Let us introduce a new symbol – the so called *curly brackets of Christoffel or brackets of Christoffel of the second kind* and define this symbol by the equation:

$$\left\{\begin{matrix} i\,k \\ l \end{matrix}\right\} = g^{l\sigma}\,\begin{bmatrix} i\,k \\ \sigma \end{bmatrix} \tag{43}$$

with the help of curly brackets of Christoffel we rewrite the equation before the last in the following way:

$$\overline{\left\{\begin{matrix} i\,k \\ s \end{matrix}\right\}} = \left\{\begin{matrix} \alpha\,\beta \\ \gamma \end{matrix}\right\}\,\frac{\partial x_\alpha}{\partial \overline{x}_i}\,\frac{\partial x_\beta}{\partial \overline{x}_k}\,\frac{\partial \overline{x}_s}{\partial x_\sigma} + \frac{\partial^2 x_\sigma}{\partial \overline{x}_i \partial \overline{x}_k}\,\frac{\partial \overline{x}_s}{\partial x_\sigma}.$$

Replacing i, k, s by $\overline{\lambda}, \overline{\mu}, \overline{i}$ we get the following equation:

$$\overline{\left\{\begin{matrix} \lambda\,\mu \\ i \end{matrix}\right\}} = \left\{\begin{matrix} \alpha\,\beta \\ \sigma \end{matrix}\right\}\,\frac{\partial x_\alpha}{\partial \overline{x}_\lambda}\,\frac{\partial x_\beta}{\partial \overline{x}_\mu}\,\frac{\partial \overline{x}_i}{\partial x_\sigma} + \frac{\partial^2 \sigma}{\partial \overline{x}_\lambda \partial \overline{x}_\mu}\,\frac{\partial \overline{x}_i}{\partial x_\sigma}. \tag{44}$$

This equation shows that Christoffel's curly brackets can be considered as tensorial parameters:

$$\Gamma^i_{\lambda\mu} = \left\{\begin{matrix} \lambda\,\mu \\ i \end{matrix}\right\}.$$

In such a way, with the help of Christoffel's curly brackets we can form tensor derivatives of any order. Unlike the general tensor derivatives we will call the tensor derivatives, obtained with the help of Christoffel's brackets – *Riemannian tensor derivatives*; we notice that the brackets of Christoffel contain a combination of components of the fundamental tensor g_{ik} and the derivatives of these components with respect to the variables $x_l : g_{ik,l}$.

It is not difficult to define, with the help of Christoffels's brackets, the general form of the tensorial parameters; subtracting equation (44) from equation (27) and denoting by $A^i_{\lambda\mu}$ the difference:

$$\Gamma^i_{\lambda\mu} - \left\{ \begin{matrix} \lambda\,\mu \\ i \end{matrix} \right\} = A^i_{\lambda\mu},$$

we will have:

$$\overline{A}^i_{\lambda\mu} = A^\sigma_{\alpha\beta} \frac{\partial x_\alpha}{\partial \overline{x}_\lambda} \frac{\partial x_\beta}{\partial \overline{x}_\mu} \frac{\partial \overline{x}_i}{\partial x_\sigma}.$$

In this way, the only restriction on $A^i_{\lambda\mu}$ is that, it should be a tensor: from here it is clear, that the general form of tensorial parameters can be expressed by the formula:

$$\Gamma^i_{\lambda\mu} = \left\{ \begin{matrix} \lambda\,\mu \\ i \end{matrix} \right\} + A^i_{\lambda\mu},$$

where $A^i_{\lambda\mu}$ is a mixed tensor of the third rank.

In the same way as relation (28) was obtained from relation (27) from equation (44) we can determine $\frac{\partial^2 x_\nu}{\partial \overline{x}_\lambda \partial \overline{x}_\mu}$ and obtain the following system of equations:

$$\frac{\partial^2 x_\nu}{\partial \overline{x}_\lambda \partial \overline{x}_\mu} = \left\{ \begin{matrix} \lambda\,\mu \\ i \end{matrix} \right\} \frac{\partial x_\nu}{\partial \overline{x}_i} - \left\{ \begin{matrix} \lambda\,\beta \\ \nu \end{matrix} \right\} \frac{\partial x_\alpha}{\partial \overline{x}_\lambda} \frac{\partial x_\beta}{\partial \overline{x}_\mu}. \tag{45}$$

Below we will see that the Christoffel symbols are very important for the geometry of curved spaces: these symbols allow us to solve, in many cases, the problem of going with the help of transformations of variables from a system of n^3 quantities g_{ik} to a system of n^2 of quantities $\overline{g}_{ik}$ – which, as we will see later, from a geometrical point of view is equivalent to whether two spaces differ significantly or their difference is only due to the different systems of coordinates used in the two spaces. To these questions we will come later, now we will introduce several simplest properties of Christoffel's brackets the greater part of which will not even need to be proved.

$$\left[\begin{matrix} i\,k \\ l \end{matrix} \right] = \left[\begin{matrix} k\,i \\ l \end{matrix} \right], \tag{46}$$

$$\left\{ \begin{matrix} i\,k \\ l \end{matrix} \right\} = \left\{ \begin{matrix} k\,i \\ l \end{matrix} \right\} \tag{47}$$

In such a way, the tensorial parameters, formed with the help of Christoffel brackets, are symmetric parameters.

$$\frac{\partial g_{ik}}{\partial x_l} = g_{ik,l} = \begin{bmatrix} i\, l \\ k \end{bmatrix} + \begin{bmatrix} k\, l \\ i \end{bmatrix} \tag{48}$$

Expressing the squire bracket of Christoffel through the curly bracket, we find:

$$\begin{bmatrix} i\, k \\ l \end{bmatrix} = g_{l\sigma} \begin{Bmatrix} i\, k \\ \sigma \end{Bmatrix}, \tag{49}$$

which equation is obtained by a multiplication of both sides of equation (43) by g_{ml} and summing them over l from 1to n and replacing m by l.

We combine equations (49) and (49) and find:

$$\frac{\partial g_{ik}}{\partial x_l} = g_{ik,l} = \begin{Bmatrix} i\, l \\ \sigma \end{Bmatrix} g_{k\sigma} + \begin{Bmatrix} k\, l \\ \sigma \end{Bmatrix} g_{\sigma i}. \tag{50}$$

This equation can be written in this way:

$$g_{ik,l} - \begin{Bmatrix} i\, l \\ \sigma \end{Bmatrix} g_{k\sigma} - \begin{Bmatrix} k\, l \\ \sigma \end{Bmatrix} g_{\sigma i} = 0,$$

and this in turn shows that the Riemannian tensor derivative of the fundamental tensor is equal to zero; so when we form Riemannian tensor derivatives the fundamental tensor plays the role of a constant.

It is more difficult to prove the formulas that will follow. Let us first prove the following equation:

$$\frac{\partial lg\,\sqrt{g}}{\partial x_l} = \begin{Bmatrix} \alpha\, l \\ \alpha \end{Bmatrix}. \tag{51}$$

Let us recall that for obtaining the derivative of a determinant each of its minors of the first order should be taken with the corresponding sign and multiply it by the derivative of the corresponding element: in such a way, according to the previous notation, we will have:

$$\frac{\partial g}{\partial x_l} = (-1)^{i+k}\, D^{ik}\, g_{ik,l}.$$

Dividing both sides of the equation by g we find:

$$\frac{1}{g} \frac{\partial g}{\partial x_l} = g^{ik}\, g_{ik,l},$$

124

but

$$g_{ik,l} = \begin{bmatrix} i\,l \\ k \end{bmatrix} + \begin{bmatrix} k\,l \\ i \end{bmatrix}$$

so:

$$\frac{1}{g}\frac{\partial g}{\partial x_l} = g^{ik}\begin{bmatrix} i\,l \\ k \end{bmatrix} + g^{ik}\begin{bmatrix} k\,l \\ i \end{bmatrix} = \left\{\begin{matrix} i\,l \\ i \end{matrix}\right\} + \left\{\begin{matrix} k\,l \\ k \end{matrix}\right\} = 2\left\{\begin{matrix} \alpha\,l \\ \alpha \end{matrix}\right\},$$

which proves formula (51).

Let us now clarify the derivatives g^{ik} with respect to x_i; introduce the following equation:

$$\frac{\partial g^{ik}}{\partial x_l} + \left\{\begin{matrix} l\,\alpha \\ i \end{matrix}\right\} g^{\alpha k} + \left\{\begin{matrix} l\,\alpha \\ k \end{matrix}\right\} g^{\alpha i} = 0. \tag{52}$$

Differentiating the following equation with respect to x_l

$$g^{ik}\,g_{km} = \delta^i_m$$

we find:

$$\frac{\partial g^{ik}}{\partial x_l}\,g_{km} + g^{ik}\frac{\partial g_{km}}{\partial x_l} = 0.$$

Using formula (50), we get:

$$\frac{\partial g^{ik}}{\partial x_l}\,g_{km} + g^{ik}\left\{\begin{matrix} k\,l \\ \sigma \end{matrix}\right\} g_{m\sigma} + g^{ik}\left\{\begin{matrix} m\,l \\ \sigma \end{matrix}\right\} g_{\sigma k} = 0.$$

Multiplying both sides of this equation by g^{ms} and summing over m from 1 to n, we will have:

$$\frac{\partial g^{ik}}{\partial x_l}\,g_{km}\,g^{ms} + g^{ik}\left\{\begin{matrix} k\,l \\ \sigma \end{matrix}\right\} g_{m\sigma}\,g^{ms} + g^{ik}\,g_{\sigma k}\left\{\begin{matrix} m\,l \\ \sigma \end{matrix}\right\} g^{ms} = 0,$$

or:

$$\frac{\partial g^{ik}}{\partial x_l}\,\delta^s_k + g^{ik}\left\{\begin{matrix} k\,l \\ \sigma \end{matrix}\right\} \delta^s_\sigma + g^{ms}\left\{\begin{matrix} m\,l \\ \sigma \end{matrix}\right\} \delta^i_\sigma = 0,$$

from where:

$$\frac{\partial g^{is}}{\partial x)_l} + g^{ik}\left\{\begin{matrix} k\,l \\ s \end{matrix}\right\} + g^{ms}\left\{\begin{matrix} m\,l \\ i \end{matrix}\right\} = 0.$$

Replacing s with k and summing the variables over α we get formula (52).

It is easy to find such a relation from formula (52) :

$$\frac{1}{\sqrt{g}}\frac{\partial g^{ik}\sqrt{g}}{\partial x_k} + \left\{\begin{matrix} \alpha\,\beta \\ i \end{matrix}\right\} g^{\alpha\beta} = 0. \tag{53}$$

Setting $l = k$ in the formula (52) and summing over k from 1to n we will have:

$$0 = \frac{\partial g^{ik}}{\partial x_k} + \begin{Bmatrix} k\,\alpha \\ i \end{Bmatrix} g^{\alpha k} + \begin{Bmatrix} k\,\alpha \\ k \end{Bmatrix} g^{\alpha i}.$$

Substituting in the last term the variable of the summation α with k and remembering, that, according to (51) $\begin{Bmatrix} k\,\alpha \\ k \end{Bmatrix} = \frac{\partial lg\,\sqrt{g}}{\partial x_\alpha}$ we easily find formula (53).

Regarding the Christoffel curly brackets as tensorial parameters we will form the curvature of these parameters and their curvature is called the *Riemann symbol of of the second kind* which symbol is defined by the following equation:

$$\{ki, \lambda\mu\} = F^{i}_{k\lambda\mu} = \frac{\partial}{\partial x_\mu} \begin{Bmatrix} k\,\lambda \\ i \end{Bmatrix} - \frac{\partial}{\partial x_\lambda} \begin{Bmatrix} k\,\mu \\ i \end{Bmatrix} +$$
$$+ \begin{Bmatrix} k\,\lambda \\ \sigma \end{Bmatrix} \begin{Bmatrix} \sigma\,\mu \\ i \end{Bmatrix} - \begin{Bmatrix} k\,\lambda \\ \sigma \end{Bmatrix} \begin{Bmatrix} k\,\lambda \\ i \end{Bmatrix}. \tag{54}$$

From the curvature we can go to the Riemann symbol F_{iklm}. It is denoted in this case by the symbol (ki, lm) and is named the *Riemann symbol of the first kind*:

$$(ki, lm) = g_{\alpha i} F^{\alpha}_{klm} = g_{\alpha i} \{k\alpha, lm\}. \tag{55}$$

The symbols of Riemann of the first and second kind are playing very important role in the geometry of curved spaces. We will introduce the basic properties of these symbols. First of all, direct calculations give the following formula:

$$(ki, lm) = \frac{1}{2} \left(\frac{\partial^2 g_{km}}{\partial x_i \partial x_l} + \frac{\partial^2 g_{il}}{\partial x_k \partial x_m} - \frac{\partial^2 g_{kl}}{\partial x_i \partial x_m} - \frac{\partial^2 g_{im}}{\partial x_k \partial x_l} \right) +$$
$$+ g^{\sigma\tau} \left(\begin{bmatrix} k\,m \\ \sigma \end{bmatrix} \begin{bmatrix} i\,l \\ \tau \end{bmatrix} - \begin{bmatrix} k\,l \\ \sigma \end{bmatrix} \begin{bmatrix} i\,m \\ \tau \end{bmatrix} \right). \tag{56}$$

Formula (55) expresses the Riemann symbol of the first kind through the Riemann symbol of the second kind; solving this formula for the Riemann symbol of the first kind we will have:

$$\{ki, \lambda\mu\} = g^{i\sigma} (k\sigma, \lambda\mu). \tag{57}$$

The general property of the curvature of tensorial parameters give us the formula:

$$\{ki, \lambda\mu\} = -\{ki, \mu\lambda\} \tag{58}$$

126

Using formula (56) we easily find the following properties of the Riemanni symbol of the first kind:

$$(ki, lm) = -(ki, ml), \tag{59}$$

$$(ki, lm) = -(ik, lm), \tag{60}$$

$$(ki, lm) = (ik, ml), \tag{61}$$

$$(ki, lm) = (lm, ki). \tag{62}$$

$$(ki, lm) + (kl, mi) + (km, il) = 0. \tag{63}$$

Using the relations (59) – (63) we show that the number N_n of the different Riemann symbols of the first kind, is definied by the formula:

$$N_n = \frac{n^2(n^2 - 1)}{12}.$$

In such way, when $n = 2$, we have $N_2 = 1$ – one Riemann symbol, when $n = 3$ $N_3 = 6$, $n = 4$, $N_4 = 20$. For the case $n = 2$ – the only Riemann symbol will be $(12, 12)$.

For our tensorial parameters we form the contraction of the Riemann symbol F_{ik}, which we will denote now by R_{ik}:

$$R_{ik} = F^{\alpha}_{i\alpha k} = \{i\alpha, \alpha k\} = g^{\alpha\sigma}(i\sigma, \alpha k).$$

Using the definition of the Riemann symbol of the second kind we find the following formula:

$$R_{ik} = \frac{\partial}{\partial x_k}\left\{{i\,\alpha}\atop{\alpha}\right\} - \frac{\partial}{\partial x_\alpha}\left\{{i\,k}\atop{\alpha}\right\} + \left\{{i\,\alpha}\atop{\sigma}\right\}\left\{{\sigma\,k}\atop{\alpha}\right\} - \left\{{i\,k}\atop{\sigma}\right\}\left\{{\sigma\,\alpha}\atop{\alpha}\right\}. \tag{64}$$

Noticing that by formula (51) $\left\{{i\,\alpha}\atop{\alpha}\right\} = \frac{\partial lg\sqrt{g}}{\partial x_i}$ and $\left\{{\sigma\,\alpha}\atop{\alpha}\right\} = \frac{\partial lg\sqrt{g}}{\partial x_\sigma}$ we will transform equation (64) in the following equation:

$$R_{ik} = \frac{\partial^2 lh\sqrt{g}}{\partial x_i \partial x_k} = \frac{1}{\sqrt{g}}\frac{\partial}{\partial x_\sigma}\left(\sqrt{g}\left\{{i\,k}\atop{\sigma}\right\}\right) + \left\{{i\,\alpha}\atop{\sigma}\right\}\left\{{k\,\sigma}\atop{\alpha}\right\}, \tag{65}$$

which equation shows that R_{ik} is symmetric with respect to lower indices. So the contraction of the Riemann symbol is contragradient symmetric tensor of the second rank.

When we study the theory of gravitation of Einstein we will see a special tensor, formed by R_{ik} and the scalar curvature $R = g^{ik}R_{ik}$. This tensor is defined by the equation:

$$R_{ik} - \frac{2}{n} g_{ik} R = T_{ik} \tag{66}$$

Conversely, we will express from equation (65) the tensor R_{ik} through T_{ik}. Multiplying this equation by g^{ik}, summing over i and k from 1 to n and denoting $g^{ik}\,T_{ik}$ by T, we will have, remembering that $g^{ik}g_{ik} = n$:

$$R - 2R = T$$

from where $R = -T$ and equation (66) can be rewritten in the following way:

$$R_{ik} = T_{ik} - \frac{2}{n}\,g_{ik}\,T. \tag{67}$$

It is not difficult to see that the scalar curvature can be written in the following form by using the definition of R_{ik}:

$$R = g^{ik}\,R_{ik} = g^{ik}\,g^{\alpha\sigma}\,(i\sigma,\alpha k). \tag{68}$$

4. We already remarked that studying the Riemann symbols can give an answer to the question when, with the help of the transformations of variables, we can go from a given system of functions g_{ik} in M_n to a given system of functions $\overline{g}_{ik}$ in $\overline{M}_n$, where these functions are connected with each other through the tensorial relation:

$$\overline{g}_{ik} = g_{\alpha\beta}\,\frac{\partial x_\alpha}{\partial \overline{x}_i}\,\frac{\partial x_\beta}{\partial \overline{x}_k}. \tag{*}$$

It is easy to see that the matter we are discussing here comes to the point that we should define from equations (*) the quantities x_i as functions of $\overline{x}_k$; as the number of equations (*) is equal to $\frac{n(n+1)}{2}$, and the number of unknown functions is n, generally speaking we should impose a lot of conditions on the quantities g_{ik} and $\overline{g}_{ik}$ in order that our task can be solved.

The geometrical significance of this task will be completely clarified in the next chapter. For now we will show that the element of the arc ds in a space of n dimensions, described with the help of the manifold M_n, will be given by the formula:

$$ds^2 = g_{ik}\,dx_i\,dx_k.$$

In another space the element of the arc $d\overline{s}$ will be defined by the equation:

$$d\overline{s}^2 = \overline{g}_{ik}\,d\overline{x}_i\,d\overline{x}_k$$

If we deal with the same space, but only described with the help of different manifolds, then of course the corresponding elements of the arc will be equal, and using g_{ik} and $\overline{x}_{ik}$ we will arrive at the formula

128

(*). The quantities g_{ik} characterize the metric of space. The solution of the task we discuss will give the answer to the question of whether g_{ik} and $\overline{g}_{ik}$ belong to the same space, described with the help different manifolds (different coordinate systems), or they belong to spaces with different metrics.

Let us prove the following theorem:

Theorem 8. *If having $\overline{x}_i = \overline{x}_i^{(0)}$ $(i = 1, 2, \ldots n)$ the following relations are satisfied*

$$\overline{g}_{ik} = g_{\alpha\beta} \frac{\partial x_\alpha}{\partial \overline{x}_i} \frac{\partial x_\beta}{\partial \overline{x}_k} \tag{*}$$

and if for all $\overline{x}_i$ the following equations hold:

$$\frac{\partial^2 x_\nu}{\partial \overline{x}_\lambda \partial \overline{x}_\mu} = \overline{\left\{ \begin{matrix} \lambda\,\mu \\ i \end{matrix} \right\}} \frac{\partial x_\nu}{\partial \overline{x}_i} - \left\{ \begin{matrix} \alpha\,\beta \\ \nu \end{matrix} \right\} \frac{\partial x_\alpha}{\partial \overline{x}_\lambda} \frac{\partial x_\beta}{\partial \overline{x}_\mu}, \tag{**}$$

then the relation (*) holds for all $\overline{x}_i$.

We express through a_{ik} the expression:

$$a_{ik} = \overline{g}_{ik} - g_{\alpha\beta} \frac{\partial x_\alpha}{\partial \overline{x}_i} \frac{\partial x_\beta}{\partial \overline{x}_k},$$

when $\overline{x}_i = \overline{x}_i^{(0)}$, $a_{ik} = 0$.

We find the derivative a_{ik} with respect to $\overline{x}_l$:

$$\frac{\partial a_{ik}}{\partial \overline{x}_l} = \frac{\partial \overline{g}_{ik}}{\partial \overline{x}_l} - \frac{\partial g_{\alpha\beta}}{\partial x_\gamma} \frac{\partial x_\alpha}{\partial \overline{x}_i} \frac{\partial x_\beta}{\partial \overline{x}_k} \frac{\partial x_\gamma}{\partial \overline{x}_l} - g_{\alpha\beta} \frac{\partial^2 x_\alpha}{\partial \overline{x}_i \partial \overline{x}_l} \frac{\partial x_\beta}{\partial \overline{x}_k} -$$

$$- g_{\alpha\beta} \frac{\partial x_\alpha}{\partial \overline{x}_i} \frac{\partial^2 x_\beta}{\partial \overline{x}_k \partial \overline{x}_l},$$

but according to equation (**) we have:

$$\frac{\partial^2 x_\alpha}{\partial \overline{x}_i \partial \overline{x}_l} = \overline{\left\{ \begin{matrix} i\,l \\ \sigma \end{matrix} \right\}} \frac{\partial x_\alpha}{\partial \overline{x}_\sigma} - \left\{ \begin{matrix} r\,s \\ \alpha \end{matrix} \right\} \frac{\partial x_r}{\partial \overline{x}_i} \frac{\partial x_s}{\partial \overline{x}_l},$$

$$\frac{\partial^2 x_\beta}{\partial \overline{x}_k \partial \overline{x}_l} = \overline{\left\{ \begin{matrix} k\,l \\ \sigma \end{matrix} \right\}} \frac{\partial x_\beta}{\partial \overline{x}_\sigma} - \left\{ \begin{matrix} r\,s \\ \beta \end{matrix} \right\} \frac{\partial x_r}{\partial \overline{x}_i} \frac{\partial x_s}{\partial \overline{x}_l}.$$

Substituting these equations in the equation above, we find after some changes of the indices:

$$\frac{\partial a_{ik}}{\partial \overline{x}_l} = \frac{\partial \overline{g}_{ik}}{\partial \overline{x}_l} - \overline{\left\{ \begin{matrix} i\,l \\ \sigma \end{matrix} \right\}} g_{\alpha\beta} \frac{\partial x_\alpha}{\partial \overline{x}_\sigma} \frac{\partial x_\beta}{\overline{x}_k} - \overline{\left\{ \begin{matrix} k\,l \\ \sigma \end{matrix} \right\}} g_{\alpha\beta} \frac{\partial x_\alpha}{\partial \overline{x}_i} \frac{\partial x_\beta}{\partial \overline{x}_\sigma} -$$

$$- \frac{\partial x_\alpha}{\partial \overline{x}_i} \frac{\partial x_\beta}{\partial \overline{x}_k} \frac{\partial x_\gamma}{\partial \overline{x}_l} \left(\frac{\partial g_{\alpha\beta}}{\partial x_\gamma} - g_{\sigma\beta} \left\{ \begin{matrix} \alpha\,\gamma \\ \sigma \end{matrix} \right\} - g_{\alpha\sigma} \left\{ \begin{matrix} \beta\,\gamma \\ \sigma \end{matrix} \right\} \right).$$

According to formula (50) the last curly bracket on the right side of this formula is equal to zero; the second and the third term are transformed in the following way:

$$g_{\alpha\beta}\frac{\partial x_\alpha}{\partial \overline{x}_\sigma}\frac{\partial x_\beta}{\partial \overline{x}_k} = \overline{g}_{\sigma k} - a_{\sigma k}, \quad g_{\alpha\beta}\frac{\partial x_\alpha}{\partial \overline{x}_i}\frac{\partial x_\beta}{\partial \overline{x}_\sigma} = \overline{g}_{i\sigma} - a_{i\sigma}.$$

Remembering that according to formula (50) we have:

$$\frac{\partial \overline{g}_{ik}}{\partial \overline{x}_l} = \overline{\left\{\begin{matrix} i\,l \\ \sigma \end{matrix}\right\}}\,\overline{g}_{\sigma k} + \overline{\left\{\begin{matrix} k\,l \\ \sigma \end{matrix}\right\}}\,\overline{g}_{i\sigma}$$

we find the following equation:

$$\frac{\partial \sigma_{ik}}{\partial \overline{x}_l} = \overline{\left\{\begin{matrix} i\,l \\ \sigma \end{matrix}\right\}}\,a_{\sigma k} + \overline{\left\{\begin{matrix} k\,l \\ \sigma \end{matrix}\right\}}\,a_{i\sigma}.$$

So , in such a way, we obtained a system of linear equations, where when $\overline{x}_i = \overline{x}_i^{\,o}$, $a_{ik} = 0$; by the property of systems of linear equations, this system of linear equations cannot have more than one solution, which coincides with the given quantities at the given initial variables. Such solution is obviously $a_{ik} = 0$. For this reason the expression a_{ik}, becoming zero when $\overline{x}_i = \overline{x}_i^{(0)}$, is equal to zero for all values of the variables, which proves the theorem.

The equations (**) derived from the equations (*), and in such a way if the equations (**) hold then it is a necessary and sufficient condition that, given the corresponding initial conditions, the system of equations (*) also holds.

The system of equations (**) can be transformed to a system of linear equations by introducing new functions $p_{\nu\lambda} = \frac{\partial x_\nu}{\partial \overline{x}_\lambda}$:

$$\begin{aligned} \frac{\partial x_\nu}{\partial \overline{x}_\lambda} &= p_{\nu\lambda}, \\ \frac{\partial p_{\nu\lambda}}{\partial \overline{x}_\mu} &= \overline{\left\{\begin{matrix} \lambda\,\mu \\ i \end{matrix}\right\}}\,p_{\nu i} - \left\{\begin{matrix} \alpha\,\beta \\ \nu \end{matrix}\right\}\,p_{\alpha\lambda}\,p_{\beta\mu}, \end{aligned} \qquad (***)$$

where the initial conditions should satisfy the equations:

$$\overline{g}_{ik} = g_{\alpha\beta}\,p_{\alpha i}\,p_{\alpha k} \quad \text{at } \overline{x}_s = \overline{x}_s^{(0)},\ s = 1,2,\ldots n. \qquad (***_1)$$

If the system (***) has a solution, then by the theorem of Cauchy-Kovalevskii this solution will be, in any case, completely determined, as long as when $\overline{x}_s = \overline{x}_s^{(0)}$ all $p_{\nu\lambda}$ and all x_i will be given, in such a way the system (***) can have as many as $n^2 + n$ constants in its

130

solution; but these constants, according to the conditions $(***_1)$, must satisfy $\frac{n(n+1)}{2}$ equations. So we will have maximum $\frac{n(n+1)}{2}$ arbitrary constants which can be present in the formulas transforming g_{ik} into $\overline{g}_{ik}$. It is clear that such calculations imply that the system $(***)$ will be a system that is *completely integrable*, i.e., the condition that it is an integrable system will not pose any additional conditions on g_{ik} and $\overline{g}_{ik}$ which are contained into the coefficients of this system.

Let us now discuss the question of integrability of the system $(***)$ or, which is the same, the equations $(**)$.

We already studied the general case of tensorial parameters with the help of theorem 7. In our case we deal with symmetric parameters, so $\gamma^i_{\lambda\mu} = 0$ and $\overline{\gamma}^i_{\lambda\mu} = 0$, and therefore the first condition of theorem 7 is satisfied.

The second conditions of this theorem, that the curvature $F^i_{k\lambda\mu}$ is a tensor, is obviously equivalent to the fact that the Riemann symbol F_{iklm} is a contragradient tensor of the fourth rank and, therefore, the following equations hold:

$$\overline{(ik,lm)} = (\alpha\beta,\gamma\delta) \frac{\partial x_\alpha}{\partial \overline{x}_i} \frac{\partial x_\beta}{\partial x_k} \frac{\partial x_\gamma}{\partial \overline{x}_l} \frac{\partial x_\delta}{\partial \overline{x}_m}. \tag{69}$$

These equations (69) are conditions for the integrability of equations $(**)$ or, which is the same, of equations $(***)$. Direct examination shows, that these equations include not only the given functions g_{ik} and $\overline{g}_{ik}$ but also functions which should be determined. In such a way, arriving at equation (69), a task which we posed at the beginning of this section cannot be considered completely solved. In many cases equations (69) degenerate into meaningless identities. Such kind of cases mean that the system $(**)$ or $(***)$ is completely integrable, and that, therefore, it is possible to find formulas of transformations of variables (also containing $\frac{n(n+1)}{2}$ arbitrary constants), transforming the quantities g_{ik} into $\overline{g}_{ik}$.

One of the cases when equations (69) are satisfied themselves, is the vanishing of all (ik,km) and all $\overline{(ik,lm)}$; since such vanishing of the quantities (ik,lm) holds with constant $\overline{g}_{ik}$ then it is clear from here, that if all Riemann symbols of the first kind (ik,lm) are equal to zero, then the transformations of variables of all g_{ik} can be brought to constants – which is an important fact, as we will see, in the geometrical studies.

Another case, when the equation (69) is satisfied, is when the Riemann symbols for M_n and $\overline{M}_n$ have the following properties:

$$(ik,lm) = c(g_{ik}\,g_{km} - g_{im}\,g_{kl}),$$

$$\overline{(ik, lm)} = c(\overline{g}_{il}\,\overline{g}_{km} - \overline{g}_{im}\,\overline{kl}),$$

where the constant c is the same in both formulas. Let us show that in this case equation (69) is satisfied; by eliminating c this equation will be rewritten in this way:

$$\overline{g}_{ik}\,\overline{g}_{km} - \overline{g}_{im}\,\overline{g}_{kl} = (g_{\alpha\gamma}\,g_{\beta\delta} - g_{\alpha\beta}\,g_{\beta\gamma})\,\frac{\partial x_\alpha}{\partial \overline{x}_i}\,\frac{\partial x_\beta}{\partial \overline{x}_k}\,\frac{\partial x_\gamma}{\partial \overline{x}_l}\,\frac{\partial x_\delta}{\partial \overline{x}_m} =$$

$$= g_{\alpha\gamma}\,\frac{\partial x_\alpha}{\partial \overline{x}_i}\,\frac{\partial x_\gamma}{\partial \overline{x}_l}\cdot g_{\beta\delta}\,\frac{\partial x_\beta}{\partial \overline{x}_k}\,\frac{\partial x_\delta}{\partial \overline{x}_l} - g_{\alpha\delta}\,\frac{\partial x_\alpha}{\partial \overline{x}_i}\,\frac{\partial x_\delta}{\partial \overline{x}_m}\cdot g_{\beta\gamma}\,\frac{\partial x_\beta}{\partial \overline{x}_k}\,\frac{\partial x_\gamma}{\partial \overline{x}_l},$$

So equation (69) will be a simple consequence of equation (*). In other words, the system (***) with initial conditions (***$_1$) will be a system of equations which are completely integrable. The last of these two cases is sugnificant for the geometry of curved spaces, the so called constant of the Riemannian curvature.

In the conclusion of this section we will discuss two examples of the calculation of the symbols introduced above. First of all, we will point out that in the case of $n = 2$ it is extremely easy to form a scalar from the only Riemann symbol of the first kind, in this case from the determinant g. This scalar is called curvature is defined in the following way:

$$I = \frac{(12, 12)}{g}.$$

In fact:

$$\overline{(12, 12)} = (\alpha\beta, \gamma\delta)\,\frac{\partial x_\alpha}{\partial \overline{x}_1}\,\frac{\partial x_\beta}{\partial \overline{x}_2}\,\frac{\partial x_\beta}{\partial \overline{x}_1}\,\frac{\partial x_\delta}{\partial \overline{x}_2},$$

is based on the properties of the symbol $(\alpha\beta, \gamma\delta)$ and bringing it to $(12, 12)$ we will have:

$$\overline{(12, 12)} = (12, 12)\begin{vmatrix} \frac{\partial x_1}{\partial \overline{x}_1} & \frac{\partial x_2}{\partial \overline{x}_1} \\ \frac{\partial x_1}{\partial \overline{x}_2} & \frac{\partial x_2}{\partial \overline{x}_2} \end{vmatrix}^2$$

At the same time

$$\overline{g} = g\begin{vmatrix} \frac{\partial x_1}{\partial \overline{x}_1} & \frac{\partial x_2}{\partial \overline{x}_1} \\ \frac{\partial x_1}{\partial \overline{x}_2} & \frac{\partial x_2}{\partial \overline{x}_2} \end{vmatrix}^2,$$

because $g = g_{11}\,g_{22} - g_{12}^2$; in our case, when $n = 2$, it follows from these two formulas that:

$$\frac{\overline{(12, 12)}}{\overline{g}} = \frac{(12, 12)}{g},$$

i.e. I is invariant. It is not difficult to see, that in the case of $n = 2$ J is the scalar curvature, found above, multiplied by $-\frac{1}{2}$. In fact, for the scalar curvature we have according to formula (68):

$$R = g^{ik} \, g^{\alpha\sigma} \, (i\sigma, \alpha k).$$

In more detailed way this equation can be written in the following way:

$$\begin{aligned}
R &= g^{ik} \left(g^{11} \, (i1, 1k) + g^{12} \, (i2, 1k) + g^{21} \, (i1, 2k) + g^{22} \, (i2, 2k) \right) = \\
&= g^{11} \left(g^{11} \, (11, 11) + g^{12} \, (12, 11) + g^{21} \, (11, 21) + g^{22} \, (12, 21) \right) \\
&+ g^{12} \left(g^{11} \, (11, 12) + g^{12} \, (12, 12) + g^{21} \, (11, 22) + g^{22} \, (12, 22) \right) + \\
&+ g^{21} \left(g^{11} \, (21, 11) + g^{12} \, (22, 11) + g^{21} \, (21, 21) + g^{22} \, (22, 21) \right) + \\
&+ g^{22} \left(g^{11} \, g^{11}(21, 12) + g^{12} \, (22, 12) + g^{21} \, (21, 22) + g^{22} \, (22, 22) \right) = \\
&= g^{11} \, g^{22} \, (12, 21) + g^{12} \, g^{12} \, (12, 12) + g^{21} \, g^{21} \, (21, 21) + g^{22} \, g^{11} \, (21, 12),
\end{aligned}$$

but:

$$(12, 21) = -(12, 12), (21, 21) = (12, 12), (21, 12) = -(12, 12),$$

Therefore:

$$R = -2 \left(g^{11} \, g^{23} - g^{12} g^{12} \right) (12, 12)$$

in our case

$$g = \begin{vmatrix} g_{11} & g_{12} \\ g_{21} & g_{22} \end{vmatrix},$$

therefore g^{ik} will be defined by the equations:

$$g^{11} = \frac{g^{22}}{g}, \quad g^{12} = -\frac{g_{21}}{g}, \quad g^{21} = -\frac{g_{12}}{g}, \quad g^{22} = \frac{g_{11}}{g}$$

so:

$$g^{11} \, g^{22} - g^{12} \, g^{12} = \frac{g_{11} \, g_{22} - -g_{12}^2}{g^2} = \frac{1}{g},$$

from where:

$$R = -2 \, \frac{(12, 12)}{g},$$

i.e.,

$$I = -\frac{1}{2} \, R.$$

The above calculations show, that in the case $n = 2$, the curvature of the tensor g_{ik}, which is used, as we will see later, in classical differential geometry, differs only by a constant multiplier from the introduced scalar curvature, which in turn, justified the name of curvature given by us to some tensors and scalars.

As a second example we will calculate the curly braces of Christoffel and the Riemann symbols for the case when the components of the fundamental tensor with different indices is equal to zero.

Introducing the notation:

$$g_{ii} = H_i^2, \quad i = 1, 2, \ldots n, \tag{70}$$

and calling H_i *Lameé coefficients*, we will have:

$$g = H_1^2 \, H_2^2 \ldots H_n^2$$
$$g^{ik} = 0, \quad i \neq k, \tag{71}$$
$$g^{ii} = \frac{1}{H_i^2}$$

$$\left\{ \begin{matrix} i \, k \\ l \end{matrix} \right\} = 0, \quad \left\{ \begin{matrix} i \, k \\ k \end{matrix} \right\} = \frac{\partial lg \, H_k}{\partial x_i}, \quad \left\{ \begin{matrix} k \, k \\ l \end{matrix} \right\} = -\frac{H_k}{h_l^2} \frac{\partial H_k}{\partial x_l}, \quad \left\{ \begin{matrix} k \, k \\ k \end{matrix} \right\} = \frac{\partial lg \, H_k}{\partial x_k}. \tag{72}$$

$$(ik, lm) = 0,$$
$$(ik, km) = H_k \left\{ \frac{\partial^2 H_k}{\partial x_i \partial x_m} - \frac{1}{H_i} \frac{\partial H_i}{\partial x_m} \frac{\partial H_k}{\partial x_i} - \frac{1}{H_m} \frac{\partial H_m}{\partial x_i} \frac{\partial H_k}{\partial x_m} \right\},$$
$$(ik, ki) = H_i \, H_k \left\{ \frac{\partial}{\partial x_i} \left(\frac{1}{H_k} \frac{\partial H_k}{\partial x_i} \right) + \frac{\partial}{\partial x_k} \left(\frac{1}{H_k} \frac{\partial H_i}{\partial x_k} \right) + \right. \tag{73}$$
$$\left. + \sum_{s=1}^{n} \frac{1}{H_i^2} \frac{\partial H_i}{\partial x_s} \frac{\partial H_k}{\partial x_s} - \frac{1}{H_i^2} \frac{\partial H_i}{\partial x_i} \frac{\partial H_k}{\partial x_i} - \frac{1}{H_k^2} \frac{\partial H_i}{\partial x_k} \frac{\partial H_k}{\partial x_k} \right\}.$$

where i, k, l, m are *different* indices, taking on the values $1, 2, \ldots n$.

5. Let us see how tensor calculus can be used in the transformation of equations into quantities, studied in physics, to the so called orthogonal curvilinear coordinates. Let x_1, x_2, x_3 be rectangular rectilinear coordinates in the Cartesian system of coordinates. It is known that the location of each point P can be defined with the help of the so called curvilinear coordinates q_1, q_2, q_3 related to x_i by the relations:

$$q_i = \omega_i(x_1, x_2, x_3)$$

which are point transformation of the manifold M_3 with elements (x_1, x_2, x_3) into the manifolds $M_3^{(q)}$ with elements (q_1, q_2, q_3); the quantities q_i play here the role of coordinates $\overline{x}_i$ and the manifold $\overline{M}_3$ in this case is denoted by the symbol $M^{(q)}{}_3$.

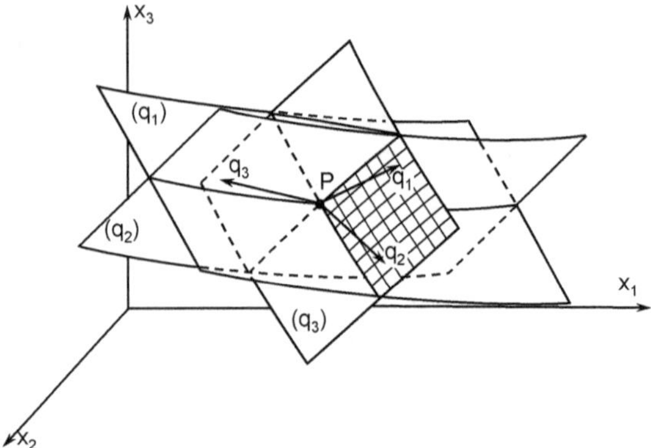

Figure 14

Geometrically the just mentioned transformation will be characterized (see Fig.14) by three systems of surfaces $q_i = const, i = 1, 2, 3$. We will call the surface $q_i = const$ the i-th *coordinate surface* and will denote it by the symbol (q_i), the intersection of the $i - th$, and the k-th coordinate surfaces which form a curve and which we will denote by the symbol $[q_i, q_k]$ and will call l *coordinate line*, where i, k, l will be three different numbers from the sequence $1, 2, 3$. At every point P we have intersection of three coordinate surfaces: $(q_1), (q_2)$ and (q_3) and three coordinate lines: $[q_2, q_3], [q_3, q_1]$ and $[q_1, q_2]$. We call the *direction* q_i at point P, the direction of the normal toward the i-th coordinate surface at point P, which is carried out in the direction of increasing values of the coordinate q_i. It is clear that the cosine of this normal with any axis x_j will be defined by the formula:

$$\cos(q_i, x_j) = \frac{1}{h_i} \frac{\partial q_i}{\partial x_j}; \quad i, j, = 1, 2, 3, \tag{74}$$

where

$$h_i = \sqrt{\left(\frac{\partial q_i}{\partial x_1}\right)^2 + \left(\frac{\partial q_i}{\partial x_2}\right)^2 + \left(\frac{\partial q_i}{\partial x_3}\right)^2}, \tag{75}$$

and we take the positive value of the squire root.

When the three directions q_1, q_2, q_3 are mutually orthogonal, then in this case we call the coordinates q_i *orthogonal curvilinear coordinates*. So in the case of orthogonal curvilinear coordinates we have the following equation:

$$\cos(q_i, q_k) = 0.$$

Then by the formula (74) we have:

$$\frac{\partial q_i}{\partial x_\sigma}\frac{\partial q_k}{\partial x_\sigma} = \frac{\partial q_i}{\partial x_1}\frac{\partial q_k}{\partial x_1} + \frac{\partial q_i}{\partial x_2}\frac{\partial q_k}{\partial x_2} + \frac{\partial q_i}{\partial x_3}\frac{\partial q_k}{\partial x_3} = 0, \qquad (76)$$

with i, k having three pairs of values: $2, 3$; $3, 1$; $1, 2$.

Multiplying both sides of the equation introduced earlier:

$$\frac{\partial x_i}{\partial q_k}\frac{\partial q_k}{\partial x_\sigma} = \delta_\sigma^i$$

by $\frac{\partial q_k}{\partial x_\sigma}$ and summing over σ using equations (75) and (76), we get:

$$h_k^2\frac{\partial x_i}{\partial q_k} = \delta_\sigma^i\frac{\partial q_k}{\partial x_\sigma} = \frac{\partial q_k}{\partial x_i} \qquad (77)$$

from where

$$h_k^4\,H_k^2 = h_k^2,$$

where:

$$H_k = \sqrt{\left(\frac{\partial x_3}{\partial q_k}\right)^2 + \left(\frac{\partial x_2}{\partial q_k}\right)^2 + \left(\frac{\partial x_3}{\partial q_k}\right)^2} \qquad (78)$$

or

$$H_k = \frac{1}{h_k}, \qquad (79)$$

$$\frac{1}{h_k}\frac{\partial q_k}{\partial q_i} = \frac{1}{H_k}\frac{\partial x_i}{\partial q_k}. \qquad (80)$$

As we saw above the quantities H_k are called Lameé coefficients and with them it is much easier to express cosine directions q with the coordinate axes for the case of orthogonal curvilinear coordinates:

$$\cos(q_i,\,x_j) = \frac{1}{H_i}\frac{\partial x_j}{\partial q_i} \qquad (81)$$

From formulas (76) and (80) we find:

$$\frac{\partial x_\sigma}{\partial q_i}\frac{\partial x_\sigma}{\partial q_k} = 0, \quad i \neq k. \qquad (82)$$

Let i, k, l be any three different numbers of the sequence of numbers $1, 2, 3$. We write equations (76) for the pair (i, l) and the pair (k, l):

$$\frac{\partial q_i}{\partial x_1}\frac{\partial q_l}{\partial x_1} + \frac{\partial q_i}{\partial x_2}\frac{\partial q_l}{\partial x_2} + \frac{\partial q_i}{\partial x_3}\frac{\partial q_l}{\partial x_3} = 0,$$

136

$$\frac{\partial q_k}{\partial x_1}\frac{\partial q_l}{\partial x_1} + \frac{\partial q_k}{\partial x_2}\frac{\partial q_l}{\partial x_2} + \frac{\partial q_k}{\partial x_3}\frac{\partial q_l}{\partial x_3} = 0.$$

This system of equations can be rewritten, as is known, in the following way:

$$\frac{\dfrac{\partial q_l}{\partial x_1}}{\dfrac{\partial q_i}{\partial x_2}\dfrac{\partial q_k}{\partial x_3} - \dfrac{\partial q_i}{\partial x_3}\dfrac{\partial q_k}{\partial x_3}} =$$

$$= \frac{\dfrac{\partial q_l}{\partial x_3}}{\dfrac{\partial q_i}{\partial x_3}\dfrac{\partial q_k}{\partial x_1} - \dfrac{\partial q_i}{\partial x_1}\dfrac{\partial q_k}{\partial x_3}} =$$

$$= \frac{\dfrac{\partial q_l}{\partial x_3}}{\dfrac{\partial q_i}{\partial x_1}\dfrac{\partial q_k}{\partial x_2} - \dfrac{\partial q_i}{\partial x_2}\dfrac{\partial q_k}{\partial x_1}}$$

Denote the common value of these three relations by λ. Squaring each of the given relations and forming a ratio, summing all numerators and all denominators of these relations we obtain for λ^2 the following expression:

$$\lambda^2 = \frac{h_l^2}{h_i^2\,h_k^2} = \frac{H_i^2\,H_k^2}{H_l^2}.$$

To find the sign of λ assume that the sequence of numbers i, k, l is formed by permutation of the signs of $1, 2, 3$; then we can prove (in the same way as in analytical geometry) that λ will be positive or negative, depending on whether the triad of the directions of q_1, q_2, q_3 has the same or different direction relative to the triad of the rectangular coordinates x_1, x_2, x_3; in other words λ will be greater than zero if it is possible to go from the triad x_1, x_2, x_3 to the triad q_1, q_2, q_3 through motion; in the opposite case λ will be smaller than zero. By taking into account all this we find:

$$\lambda = \varepsilon\,\frac{h_l}{h_i h_k} = \varepsilon\,\frac{H_i H_k}{H_l}$$

where $\varepsilon = \pm 1$. The previous three relations can be written in one formula:

$$\frac{1}{h_l}\frac{\partial q_l}{\partial x_{l'}} = \frac{\varepsilon}{h_i h_k}\left(\frac{\partial q_i}{\partial x_{i'}}\frac{\partial q_k}{\partial x_{k'}} - \frac{\partial q_i}{\partial x_{k'}}\frac{\partial q_i}{\partial x_{i'}}\right) \tag{83}$$

or

$$\frac{1}{H_l}\frac{\partial x_{l'}}{\partial q_l} = \frac{\varepsilon}{H_i h_k}\left(\frac{\partial x_{i'}}{\partial q_i}\frac{\partial x_{k'}}{\partial q_k} - \frac{\partial x_{k'}}{\partial q_i}\frac{\partial x_{i'}}{\partial q_k}\right) \tag{84}$$

where i, k, l and i', k', l' are formed from the numbers $1, 2, 3$ by permutation of the letters.

From formulas (83) and (84) it is not difficult to obtain the following equation:

$$\left\| \frac{\partial q_i}{\partial x_j} \right\| = \left\| \frac{\partial x_i}{\partial q_j} \right\| = \varepsilon$$

serving as a criterion for finding the values of ε and therefore for finding whether the triad of directions q_1, q_2, q_3 has the same or different direction with respect to the triad of the coordinate axes.

Lameé coefficients H_i are part of the components of some tensor for M_3 (q), where this tensor is easier to define in M_3. We choose for the fundamental tensorg_{ik} such a tensor whose components in M_3 will be defined by the formulas:

$$g_{ii} = 1, g_{ik} = 0, i \neq k \tag{*}$$

so, for M_3 we will have:

$$ds^2 = dx_1^2 + d_2^2 + d_3^2.$$

In our system of rectangular coordinates ds will be a length element of the ark of the curve. Let us form the components of the chosen tensor in $M_3^{(q)}$ (i.e in $\overline{M_3}$):

$$\overline{g}_{ik} = g_{\alpha\beta} \frac{\partial x_\alpha}{\partial q_i} \frac{\partial x_\beta}{\partial q_k} = \frac{\partial x_\alpha}{\partial q_i} \frac{\partial x_\alpha}{\partial q_k}.$$

In other words, using the equations (*) and (78) we have

$$\overline{g}_{ii} = H_i^2, \quad \overline{g}_{ik} = 0, \quad i \neq k.$$

In such a way the quadratic Lameé coefficients will be three of the components of the introduced fundamental tensor in $M_3^{(q)}$, and we will have:

$$ds^2 = dx_1^2 + dx_2^2 + dx_3^2 = H_1^2\, dq_1^2 + H_2^2\, dq_2^2 + H_2^3\, dq_3^2$$

If we denote the cogradient and contragradient vectors in M_3 with a directed line in the system of rectangular coordinates, whose projection on the coordinate axes will be its components then we can pose the question of finding the projection of the vector on the direction q_i in the system of orthogonal curvilinear coordinates. We will denote the projection of cogradient vector $\mathbf{a}^i$ on the direction q_i by the symbol $a^{(qi)}$ and the projection of contragradient vector $\mathbf{b}_i$ on the direction q_i will be denoted by the symbol b_q; from the theory of projection we will have:

$$a^{(q_i)} = a^s \, \cos(x_s, q_i) = \frac{1}{h_i} a^s \frac{\partial q_i}{\partial x_s},$$

138

$$b_q = b_s \, \cos(x_s, q_i) = \frac{1}{H_i} \, b_s \, \frac{\partial x_s}{\partial q_i},$$

but according to the definition of vectors, we will have;

$$\overline{a}^i = a^s \, \frac{\partial q_l}{\partial x_s}, \quad \overline{b}_i = b_s \, \frac{\partial x_s}{\partial q_l}$$

where $\overline{a}_i$ and $\overline{b}_i$, are components of the vectors in $M_i^{(q)}$; taking this in to account, from the previous formulas we derive:

$$a^{(q_i)} = \frac{1}{h_i}, \overline{a}^i = H_i \, \overline{a}^i, \quad \overline{a}^i = h_i \, a^{(q_i)} = \frac{1}{H_i} \, a^{(q_i)},$$

$$b_q = \frac{1}{H_i} \, \overline{b}_i = h_i \, \overline{b}_i, \quad \overline{b}_i = H_i \, b_{q_i} = \frac{1}{h_i} \, b_{q_i} \tag{85}$$

In such a way, the projections of the vectors on the axes q_i *do not coincide*, generally speaking, with the components of the vectors in the manifold $M_3^{(q)}$, but are different from them by multipliers depending on th Lamée coefficients. In many physical situations, as we will see later, these projections of vectors on the axes q_i have given physical meaning; therefore it is more useful to consider namely these projections, and not the components of vectors in $M_3^{(q)}$.

Consider as the simplest examples two vectors: velocity with components $\mathfrak{v}^i = \frac{dx_i}{dt}$ and a gradient with components $\mathfrak{G}_i = \frac{df}{dx_i}$. According to the discussion in §1 we will have:

$$\overline{\mathfrak{v}}^i = \frac{d\,q_i}{dt}, \quad \mathfrak{G}_i = \frac{\partial q}{\partial q_i}$$

and then from formulas (85) we find:

$$\mathfrak{v}_i^{(q_i)} = H_i \, \frac{d\,q_i}{dt}, \quad \mathfrak{G}_q = \frac{1}{H_i} \, \frac{\partial f}{\partial q_i}.$$

In such way tensor calculus makes it possible to obtain immediately the projections of the velocity and gradient on the directions q_i connected with the system of orthogonal curvilinear coordinates.

We know that in M_3 an antisymmetric tensor of the second rank (cogradient or contragradient) has only three different from one another and different from zero components. We will call *cogradient pseudo-vector* the set of three quantities: a^1, a^2, a^3, defined for each of the manifolds M_3 by the equations:

$$a^1 = T^{23} = -T^{32}, \quad a^2 = T^{31} = -T^{13}, \quad a^2 = T^{12} = -T^{21},$$

where T^i is an antisymmetric cogradient tensor; also the set of three quantities $a_1, a_2.a_3$ are defined for each of M_3 by the equations:

$$a_1 = T_{23} = -T_{32} \quad a_2 = T_{31} = -T_{13}; \quad a_3 = T_{12} = -T_{21},$$

where $\mathbf{T}_{ik}$ is an antisymmetric tensor, will be called *contragradient pseudo-vector*.

We will show first of all, that the introduced pseudo-vectors will be vectors of the subgroup of motion of the affine orthogonal group in M_3. As cogradient and contragradient tensors of the affine orthogonal group (see §2) do not differ from each other, then to prove our assertion, it is sufficient to consider one of the pseudo-vectors; we will consider the case of a contragradient tensor. Remembering, that according §1, for affine orthogonal transformations we will have:

$$\frac{\partial x_l}{\partial \overline{x}_i} = a_l^i,$$

and noticing that a contragredient tensor is transformed by the formula:

$$\overline{T}_{ik} = T_{s\sigma} \frac{\partial x_s}{\partial \overline{x}_i} \frac{\partial x_\sigma}{\partial \overline{x}_k},$$

we find

$$\overline{T}_{ik} = T_{s\sigma} \, \alpha_s^i \, \alpha_\sigma^k$$

but $T_{s\sigma} = 0$ at $s = \sigma$, and $T_{s\sigma} = -T_{\sigma s}$ so using the definition of a pseudo-vector we find:

$$\overline{T}_{ik} = a_1 \, (\alpha_2^i \, \alpha_3^k - \alpha_3^i \, \alpha_2^k) + a_2 \, (\alpha_3^i \, \alpha_1^k - \alpha_1^i \, \alpha_3^k) + (\alpha_1^i \, \alpha_2^k - \alpha_2^i \, \alpha_1^k);$$

but for the transformations of the subgroup of motion we will have:

$$\alpha_2^i \, \alpha_3^k - \alpha_3^i \, \alpha_2^k = \alpha_1^l, \quad \alpha_3^i \, \alpha_1^k - \alpha_1^i \, \alpha_3^k = \alpha_2^l, \quad \alpha_1^i \, \alpha_2^k - \alpha_2^i \, \alpha_1^k = \alpha_3^l,$$

where i, k, l form one of the sequences of indices $(2,3,1)$, $(3,1,2)$, $(1,2,3)$. So from the formula for $\overline{T}_{ik}$, we find:

$$\overline{a}_1 = \overline{T}_{23} = \alpha_1^1 \, a_1 + \alpha_2^1 \, a_2 + \alpha_3^1 \, a_2,$$
$$\overline{a}_1 = \overline{T}_{31} = \alpha_1^2 \, a_1 + \alpha_2^2 \, a_2 + \alpha_3^2 \, a_3,$$
$$\overline{a}_3 = \overline{T}_{12} = \alpha_1^3 \, a_1 + \alpha_2^3 \, a_2 + \alpha_2^2 \, a_3.$$

These formulas show that for each of the manifold M_3, the three quantities a_1, a_2, a_3 define a vector for the subgroup of motion of the

140

affine orthogonal group. It is not difficult to clarify with the analogous considerations that this vector is the so-called *axial vector* whose components are formed through reflection.

The vector product of two vectors (cogradient or contragradient) and also the curl of contragradient vector, are typical pseudo-vectors. Let us see how, with the help of tensor calculus, we calculate the projections of pseudo-vectors on the directions q_i of a system of orthogonal curvilinear coordinates. Here, of course, under projection of a pseudo-vector we understand the projection of a directed segment whose projections on the axes of a rectangular coordinates will be the components of our pseudo-vector. We denote the projection on the direction q_i of the cogradient pseudo-vector $\mathfrak{a}^i$ and the contragradient pseudo-vector $\mathfrak{b}_i$ by the symbols $\mathfrak{a}^{(q^i)}$ and $\mathfrak{b}_{q_i}$

Let the cogradient pseudo-vector $\mathfrak{a}^i$ be formed, with the help of an antisymmetric tensor $\mathbf{T}^{ik}$, and see how $\mathfrak{a}^{(q_i)}$ will be connected with $\overline{\mathfrak{a}}^i$ components of our pseudo-vector in $M^{(q)}$. We have:

$$\overline{T}^{ik} = T^{s\sigma} \frac{\partial q_i}{\partial x_s} \frac{\partial q_k}{\partial x_\sigma} = T^{23}\left(\frac{\partial q_i}{\partial x_2} \frac{\partial q_k}{\partial x_3} - \frac{\partial q_i}{\partial x_3} \frac{\partial q_k}{\partial x_2} \right) +$$

$$+ T^{31}\left(\frac{\partial q_i}{\partial x_2} \frac{\partial q_k}{\partial x_1} - \frac{\partial q_i}{\partial x_1} \frac{\partial q_i}{\partial x_2} \right) + T^{12}\left(\frac{\partial q_1}{\partial x_1} \frac{\partial q_k}{\partial x_2} - \frac{\partial q_k}{\partial x_2} \frac{\partial q_k}{\partial x_1} \right),$$

from where, using equations (83), we find;

$$\overline{T}^{ik} = \varepsilon \frac{h_i h_k}{h_l}\left(a^1 \frac{\partial q_l}{\partial x_1} + a^2 \frac{\partial q_l}{\partial x_2} + a^3 \frac{\partial q_l}{\partial x_2} \right).$$

But, by definition, we will have:

$$\overline{T}^{ik} = \overline{a}^l, \qquad \frac{1}{h_l} a^1 \frac{\partial q_l}{\partial x_1} + a^2 \frac{\partial q_l}{\partial x_2} + a^3 \frac{\partial q_l}{\partial x_2} = a^{(q_l)},$$

from where:

$$\overline{a}^l = \varepsilon\, h_i\, h_k\, a^{(q)l},$$

where (i, k, l) will be one of the three numbers (2,3,1), (3,1,2), (1,2,3).

In the same way, for the contragradient pseudo-vector we can create the following table of formulas, analogous to formulas (85):

$$a^{(q_l)} = \frac{\varepsilon}{h_i h_k} \overline{a}^l - \varepsilon\, H_i\, H_k\, \overline{a}^l, \qquad \overline{a}^l - \frac{\varepsilon}{H_i H_k} a^{(q_l)} = \varepsilon\, h_i\, h_k\, a^{(q_l)},$$

$$a_{q_l} = \varepsilon\, h_i\, h_k\, \overline{a}_l = \frac{\varepsilon}{H_i H_k} \overline{a}_l, \qquad \overline{a}_l = \varepsilon\, H_i\, H_k\, a_{q_l} = \frac{\varepsilon}{h_i h_k} a_{q_l},$$

$$(86)$$

where (i, k, l) will be one of three numbers (2,3,1), (3,1,2), (1,2,3), but $\varepsilon = \pm 1$ in accordance with the explanation above.

We will apply the previous methods to the transformation of the vector product. We introduce, with the help of the vectors $\mathbf{a}^i, \mathbf{b}^i, \mathbf{a}_i, \mathbf{b}_i$, the following two antisymmetric tensors:

$$T^{ik} = a^i\, b^k - a^k\, b^i,$$
$$T_{ik} = a_i\, b_k - a_k\, b_i$$

and form from these tensors two pseudo-vectors with components:

$$\mathfrak{a}^1 = a^2\, b^2 - a^3\, b^3, \quad \mathfrak{a}^2 = a^3\, b^1 - a^1\, b^3, \quad \mathfrak{a}^3 = a^1\, b_2 - a_2\, b^1,$$

$$\mathfrak{b}_1 = a_2\, b_3 - a_3\, b_2, \quad \mathfrak{b}_2 = a_3\, b_1 - a_1\, b_3, \quad \mathfrak{b}_3 = a_1\, b_2 - a_2\, b_1;$$

These pseudo-vectors are called vector products of cogradient vectors $\mathbf{a}^i, \mathbf{b}^i$ or of the corresponding contragradient vectors $\mathbf{a}_i, \mathbf{b}_i$.

By formula (86), we will have:

$$\mathfrak{b}_{q1} = \frac{\varepsilon}{H_2 H_3}\, \overline{\mathfrak{b}}_1 = \frac{\varepsilon}{H_2 H_3}\, (\overline{a}_2 \overline{b}_3 - \overline{a}_3 \overline{b}_2).$$

But from formula (85) we find:

$$\overline{a}_2 = H_2\, a_{q_2}, \quad \overline{a}_3 = H_3\, a_{q_3}, \overline{b}_2 = H_2\, b_{q_2}, \quad \overline{b}_3 = H_3\, b_{q_3},$$

from where:

$$\begin{aligned}
\mathfrak{b}_{q_1} &= \varepsilon\, (a_{q_2}\, b_{q_3} - a_{q_3}\, b_{q_2}),\\
\mathfrak{b}_{q_2} &= \varepsilon\, (a_{q_3}\, b_{q_1} - a_{q_1}\, b_{q_3}),\\
\mathfrak{b}_{q_3} &= \varepsilon\, (a_{q_1}\, b_{q_2} - a_{q_2}\, b_{q_1})
\end{aligned} \tag{87}$$

and analogous formulas for the vector product of cogradient vectors can be found. The quantity $\varepsilon = +1$ as long as the triad of directions q_1, q_2, q_3 has the same direction as the triad of the coordinate axes of our rectangular coordinate system.

Consider further the transformation of the pseudo-vector of curl. We will define, with the help with contragradient vector $\mathbf{a}_i$, for each set of the following numbers in M_3:

$$T_{ik} = \frac{\partial a_i}{\partial x_i} - \frac{\partial a_i}{\partial x_k}.$$

In the next section we will prove that T_{ik} are components of some antisymmetric contragradient tensor. With the help of this tensor we form a contragradient pseudo-vector $\mathfrak{c}_i$:

$$\mathfrak{c}_1 = \frac{\partial a_3}{\partial x_2} - \frac{\partial a_2}{\partial x_3}, \quad \mathfrak{c}_2 = \frac{\partial a_1}{\partial x_3} - \frac{\partial a_3}{\partial x_1}, \quad \mathfrak{c}_3 = \frac{\partial a_2}{\partial x_1} - \frac{\partial a_1}{\partial x_2},$$

142

This pseudo-vector is called the *curl of the vector a_i in M_3*. Using formulas (86), will have:

$$c_{q_1} = \frac{\varepsilon}{H_2\,H_3}\,\bar{c}_1$$

but $\bar{c}_1 = \frac{\partial \bar{a}_2}{\partial q_2} - \frac{\partial \bar{a}_2}{\partial q_3}$, and on the other hand by formula (85) we will have:

$$\bar{a}_3 = H_3\,a_{q_3}, \quad \bar{a}_2 = H_2\,a_{q_2},$$

because the projection of the curl on the directions q_1, q_2, q_3 will be:

$$c_{q_1} = \frac{\varepsilon}{H_2\,H_3}\left(\frac{\partial H_3\,a_{q_3}}{\partial q_2} - \frac{\partial H_2\,a_{q_2}}{\partial q_2} \right),$$

$$c_{q_2} = \frac{\varepsilon}{H_2\,H_3}\left(\frac{\partial H_1\,a_{q_1}}{\partial q_3} - \frac{\partial H_3\,a_{q_3}}{\partial q_1} \right), \qquad (88)$$

$$c_{q_3} = \frac{\varepsilon}{H_1\,H_2}\left(\frac{\partial H_2 a_{q_2}}{\partial q_1} - \frac{\partial H_1\,a_{q_1}}{\partial q_2} \right).$$

Let us apply the obtained formulas for the so called cylindrical coordinate system defined by the formulas:

$$x_1 = q_1 \cos q_2, \quad x_2 = q_1 \sin q_2, \quad x_3 = q_3,$$

or, in more usual notations:

$$x_1 = r \cos\theta, \quad x_2 = r \sin\theta, \quad x_3 = z.$$

Forming Lamé coefficients we will have:

$$H_1 = 1, \quad H_2 = q_1 = r, \quad H_3 = 1,$$

Forming the determinant $\left\| \frac{\partial x_i}{\partial q_j} \right\|$ we find that it will be equal to $+1$, because $\varepsilon = +1$: formulas (87) and (88) will become:

$$b_r = a_\theta\,b_z - a_z\,b_\theta, \quad b_\theta = a_z\,b_r - a_r\,b_z, \quad b_z = a_r\,b_\theta - a_\theta\,b_r,$$

$$c_r = \frac{1}{r}\frac{\partial a_z}{\partial \theta} - \frac{\partial a_\theta}{\partial z}, \quad c_\theta = \frac{\partial a_r}{\partial z} - \frac{\partial a_z}{\partial r}, \quad c_z = \frac{\partial a_\theta}{\partial r} - \frac{\partial a_r}{\partial \theta} + \frac{a_\theta}{r}.$$

If it is necessary to transform some expressions into curvilinear coordinates, having derivatives of vectors and tensors with respect to the coordinates, we do that in the following way: we replace the ordinary tensor derivatives and, using the formulas of transformations of tensors and tensorial parameters, we get the transformation for the given expression in the case of curvinear coordinates.

In conclusion we will remark that the Lamé coefficients cannot be defined arbitrarily. In fact, these coefficients are part of the components in $M_3^{(q)}$ of the fundamental tensor, which in M_3 has its own components of constant numbers. In M_3 the Riemann symbols of the first kind for our fundamental tensor are equal to zero, according to the discussion in the previous section, the Riemann symbols of the first kind should be equal to zero for the same fundamental tensor in $M_3^{(q)}$; these symbols (which are different in the sense mentioned in the previous section) will be $N_3 = 6$; so we will get six relations, called Lam]'e equations. These six relations can be written as:

$$(21, 13) = 0, \quad (32, 21) = 0, \quad (12, 32) = 0,$$

$$(12, 12) = 0, \quad (23, 23) = 0, \quad (31, 31) = 0$$

From where, using equations (73), we write the Lamé equations in the following form:

$$\frac{\partial^2 H_1}{\partial q_3 \partial q_2} - \frac{1}{H_2} \frac{\partial H_2}{\partial q_3} \frac{\partial H_1}{\partial q_3} - \frac{1}{H_3} \frac{\partial H_3}{\partial q_2} \frac{\partial H_1}{\partial q_3} = 0.$$

$$\frac{\partial^2 H_2}{\partial q_3 \partial q_1} - \frac{1}{H_3} \frac{\partial H_2}{\partial q_1} \frac{\partial H_2}{\partial q_3} - \frac{1}{H_2} \frac{\partial H_1}{\partial q_3} \frac{\partial h_2}{\partial q_2} = 0.$$

$$\frac{\partial^2 H_2}{\partial q_1 \partial q_2} - \frac{1}{H_1} \frac{\partial H_1}{\partial q_2} \frac{\partial H_3}{\partial q_3} - \frac{1}{H_2} \frac{\partial H_3}{\partial q_1} \frac{\partial H_2}{\partial q_2} = 0.$$

$$\frac{\partial}{\partial q_2}\left(\frac{1}{H_2} \frac{\partial H_3}{\partial q_2}\right) + \frac{\partial}{\partial q_3}\left(\frac{1}{H_3} \frac{\partial H_2}{\partial q_3}\right) + \frac{1}{H_1^2} \frac{\partial H_2}{\partial q_1} \frac{\partial H_3}{\partial q_3} = 0.$$

$$\frac{\partial}{\partial q_3}\left(\frac{1}{H_3} \frac{\partial H_1}{\partial q_3}\right) + \frac{\partial}{\partial q_1}\left(\frac{1}{H_1} \frac{\partial H_3}{\partial q_1}\right) + \frac{1}{H_2^2} \frac{\partial H_3}{\partial q_2} \frac{\partial H_1}{\partial q_2} = 0.$$

$$\frac{\partial}{\partial q_1}\left(\frac{1}{H_1} \frac{\partial H_2}{\partial q_3}\right) + \frac{\partial}{\partial q_3}\left(\frac{1}{H_2} \frac{\partial H_1}{\partial q_2}\right) + \frac{1}{H_3^2} \frac{\partial H_1}{\partial q_3} \frac{\partial H_2}{\partial q_2} = 0.$$

6. In conclusion of the present section we will point out several very simple differential operations with tensors, which have tensorial character.

Above we saw that from functions of the variables x_i it is possible to form contragradient vector $\mathbf{f}_i$ with the help of the following equations:

$$f_i = \frac{\partial f}{\partial x_i}.$$

144

This contragradient vector $\mathbf{f}_i$ is called, as we already discussed, *gradient* and is denoted in the following way:

$$\mathbf{f}_i = \operatorname{grad} f, \quad f_i = \frac{\partial f}{\partial x_i} \tag{89}$$

Also, from any contragradient vector φ_i it is possible to form some contragradient tensor φ_{ik}, called *the curl of the vector* φ_i and defined by the following equation:

$$\varphi_{ik} = \operatorname{curl} \varphi_i, \quad \varphi_{ik} = \frac{\partial \varphi_i}{\partial x_k} - \frac{\partial \varphi_k}{\partial x_i} \tag{90}$$

Let us prove, that φ_{ik} is indeed a tensor.
We have:

$$\overline{\varphi}_{ik} = \frac{\partial \overline{\varphi}_i}{\partial \overline{x}_k} - \frac{\partial \overline{\varphi}_k}{\partial \overline{x}_i}, \quad \overline{\varphi}_i = \varphi_\alpha \frac{\partial x_\alpha}{\partial \overline{x}_i}, \quad \overline{\varphi}_k = \varphi_\alpha \frac{\partial x_\alpha}{\partial \overline{x}_k},$$

therefore

$$\overline{\varphi}_{ik} = \frac{\partial \varphi_\alpha}{\partial x_\beta} \frac{\partial x_z}{\partial \overline{x}_i} \frac{\partial x_\beta}{\partial \overline{x}_k} + \varphi_\alpha \frac{\partial^2 x_\alpha}{\partial \overline{x}_i \partial \overline{x}_k} - \frac{\partial \varphi_\alpha}{\partial x_\beta} \frac{\partial x_z}{\partial \overline{x}_k} \frac{\partial x_\beta}{\partial \overline{x}_i} - \varphi_\alpha \frac{\partial^2 x_\alpha}{\partial \overline{x}_k \partial \overline{x}_i} =$$

$$= \varphi_{\alpha\beta} \frac{\partial x_\alpha}{\overline{\partial x_i}} \frac{\partial x_\beta}{\partial \overline{x}_k},$$

from where it follows that φ_{ik} is a contragradient tensor.

It is not difficult to see that *the necessary and sufficient condition for φ_i to be a gradient is when the curl of φ_i is equal to zero*. The fact that this condition is necessary is clear by itself; but its sufficiency comes from the following considerations: if $\varphi_{ik} = \frac{\partial \varphi_i}{\partial x_k} - \frac{\partial \varphi_k}{\partial x_i}$ equals to zero, then this means that φ_i is a derivative of some functions with respect to x_i, i.e. φ_{ik} is gradient. We point out that the vector, which is gradient, is called *vector potential*.

With the help of any antisymmetric contragradient tensor of the second rank ψ_{ik} we can form a special tensor of the third rank ψ_{ikl} called cycle ψ_{ik} and defined by the following equation:

$$\psi_{ikl} = \text{Ц}\,\psi_{ik}, \quad \psi_{ikl} = \frac{\partial \psi_{ik}}{\partial x_l} + \frac{\partial \psi_{kl}}{\partial x_i} + \frac{\partial \psi_{il}}{\partial x_k}. \tag{91}$$

We will show that the cycle is a contragradient tensor. We have:

$$\overline{\psi}_{ik} = \psi_{\alpha\beta} \frac{\partial x_\alpha}{\partial \overline{x}_i} \frac{\partial x_\beta}{\partial \overline{x}_k},$$

$$\frac{\partial \overline{\psi}_{ik}}{\partial \overline{x}_l} = \frac{\partial \psi_{\alpha\beta}}{\partial x_\gamma} \frac{\partial x_\alpha}{\partial \overline{x}_i} \frac{\partial x_\beta}{\partial \overline{x}_k} \frac{\partial x_\gamma}{\partial \overline{x}_l} + \psi_{\alpha\beta} \frac{\partial^2 x_\alpha}{\partial \overline{x}_i \overline{x}_l} \frac{\partial x_\beta}{\partial \overline{x}_k} + \psi_{\alpha\beta} \frac{\partial x_\alpha}{\partial \overline{x}_i} \frac{\partial^2 x_\beta}{\partial \overline{x}_k \overline{x}_l},$$

from where

$$\overline{\psi}_{ikl} = \psi_{\alpha\beta\gamma} \frac{\partial x_\alpha}{\partial \overline{x}_i} \frac{\partial x_\beta}{\partial \overline{x}_k} \frac{\partial x_\gamma}{\partial \overline{x}_l} + \psi_{\alpha\beta} \frac{\partial^2 x_\alpha}{\partial \overline{x}_i \overline{x}_l} \frac{\partial x_\beta}{\partial \overline{x}_k} + \psi_{\alpha\beta} \frac{\partial x_\alpha}{\partial \overline{x}_i} \frac{\partial^2 x_\beta}{\partial \overline{x}_k \overline{x}_l} +$$

$$+ \psi_{\alpha\beta} \frac{\partial^2 x_\alpha}{\partial \overline{x}_k \overline{x}_i} \frac{\partial x_\beta}{\partial \overline{x}_l} + \psi_{\alpha\beta} \frac{\partial x_\alpha}{\partial \overline{x}_k} \frac{\partial^2 x_\beta}{\partial \overline{x}_l \overline{x}_i} +$$

$$+ \psi_{\alpha\beta} \frac{\partial^2 x_\alpha}{\partial \overline{x}_l \overline{x}_k} \frac{\partial x_\beta}{\partial \overline{x}_i} + \psi_{\alpha\beta} \frac{\partial x_\alpha}{\partial \overline{x}_l} \frac{\partial^2 x_\beta}{\partial \overline{x}_i \overline{x}_k} +$$

$$= \psi_{\alpha\beta\gamma} \frac{\partial x_\alpha}{\partial \overline{x}_i} \frac{\partial x_\beta}{\partial \overline{x}_k} \frac{\partial x_\gamma}{\partial \overline{x}_l} + \left(\psi_{\alpha\beta} + \psi_{\beta\alpha}\right) \frac{\partial x_\alpha}{\partial \overline{x}_i} \frac{\partial^2 x_\beta}{\partial \overline{x}_k \overline{x}_l} +$$

$$+ \left(\psi_{\alpha\beta} + \psi_{\beta\alpha}\right) \frac{\partial x_\alpha}{\partial \overline{x}_k} \frac{\partial^2 x_\beta}{\partial \overline{x}_l \overline{x}_i} + \left(\psi_{\alpha\beta} + \psi_{\beta\alpha}\right) \frac{\partial x_\alpha}{\partial \overline{x}_l} \frac{\partial^2 x_\beta}{\partial \overline{x}_i \overline{x}_k} =$$

$$= \psi_{\alpha\beta\gamma} \frac{\partial x_\alpha}{\partial \overline{x}_i} \frac{\partial x_\beta}{\partial \overline{x}_k} \frac{\partial x_\gamma}{\partial \overline{x}_l}.$$

Due to the antisymmetrical nature of the tensor ψ_{ik} (meaning that $\psi_{\alpha\beta} + \psi_{\beta\alpha} = 0$); the last equation shows that ψ_{ikl} is indeed a tensor.

It is not difficult to clarify, with the help of the notion of a cycle, the necessary and sufficient conditions in order that the given tensor would be a curl of a the contragradient vector. We will name the tensor which a curl of a contragradient vector *solenoidal tensor*.

In order that a contragradient tensor of the second rank is a solenoidal tensor, it is necessary and sufficient for it to be antisymmetric and its cycle is equal to zero.

The necessary condition is checked directly. We will find its sufficiency; obviously it is necessary, given the above conditions, to choose n for function of φ_i satisfying the equations:

$$\frac{\partial \varphi_i}{\partial x_k} - \frac{\partial \varphi_k}{\partial x_i} = \psi_{ik} \; (i, k, = 1, 2, \ldots n).$$

Since ψ_{ik} by definition is an antisymmetric tensor, then half of these equations vanish and we can write the previous system in the following form:

$$\frac{\partial \varphi_i}{\partial x_k} = \frac{\partial \varphi_k}{\partial x_i} + \psi_{ik}, \quad i = 1, 2, \ldots n, \quad k = i + 1, i + 2, \ldots n,$$

We select from these equations those which contain φ_1 and then the previous equations divide into two groups:

$$\frac{\partial \varphi_1}{\partial x_k} = \frac{\partial \varphi_k}{\partial x_1} + \psi_{1k} \quad k = 2, 3 \ldots n, \tag{*}$$

146

$$\frac{\partial \varphi_i}{\partial x_k} = \frac{\partial \varphi_k}{\partial x_i} + \psi_{ik}, \quad i = 2, 3, \ldots, n, \quad k = i+1, i+2, \ldots n, \quad (**)$$

If the system $(**)$ will have a solution, then the possibility to determine φ_1 from the equations $(*)$ will depend on the following conditions:

$$\frac{\partial^2 \varphi_1}{\partial x_k \partial x_l} = \frac{\partial^2 \varphi_k}{\partial x_1 \partial x_l} + \frac{\partial \varphi_{1k}}{\partial x_l} = \frac{\partial^2 \varphi_1}{\partial x_l \partial x_k} = \frac{\partial^2 \varphi_l}{\partial x_1 \partial x_l} + \frac{\partial \varphi_{1l}}{\partial x_k},$$

or

$$\frac{\partial}{\partial x_1} \left(\frac{\partial \varphi_k}{\partial x_l} - \frac{\partial \varphi_l}{\partial x_k} \right) + \frac{\partial \varphi_{1k}}{\partial x_l} - \frac{\partial \varphi_{1l}}{\partial x_k} = 0,$$

but from equations $(**)$ we find:

$$\frac{\partial \varphi_k}{\partial x_l} - \frac{\partial \varphi_l}{\partial x_k} = \psi_{kl}.$$

But $\psi_{1l} = -\psi_{l1}$, so the previous condition can be written in this way:

$$\frac{\partial \varphi_{kl}}{\partial x_1} = \frac{\partial \varphi_{l1}}{\partial x_k} + \frac{\partial \varphi_{1k}}{\partial x_l} = \psi_{kl1} = 0.$$

In such a way to determine φ_1 it is necessary that the equations $(**)$ had a solution and that $\psi_{kl1} = 0$. In the same way we will find how to determine $\varphi_2, \varphi_3 \ldots$ as long as ψ_{kl2}, ψ_{kl3} and so on will be equal to zero. In a similar way we will prove our assumption.

It is clear that the notions gradient and curl are direct generalizations of the classical representations in the case of tensor analysis; the notion of cycle can be regarded as the known generalization of the classical notion of divergence. Further generalization of the notion of divergence will be discussed in the next chapter.

5 INTEGRAL THEOREMS OF TENSOR CALCULUS

1. In this chapter, with the help of the notion n-fold curvilinear integral, we will generalize the known theorem of Gauss, concerning the transformation of a volume integral into a surface integral.

For the consideration of this theorem it is necessary to generalize the notion of *divergence* of a vector for the case of tensor calculus.

Consider the formation of divergence in the ordinary vector calculus; according to the definition we have the following equation:

$$\operatorname{div} \mathfrak{a} = \frac{\partial \mathfrak{a}_1}{\partial x_1} + \frac{\partial \mathfrak{a}_2}{\partial x_2} + \frac{\partial \mathfrak{a}_3}{\partial x_3} = \frac{\partial \mathfrak{a}^1}{\partial x_1} + \frac{\partial \mathfrak{a}^2}{\partial x_2} + \frac{\partial \mathfrak{a}^2}{\partial x_3},$$

because for affine vectors (and therefore in the ordinary vector analysis) there does not exist a difference between cogradient and contragradient vectors. Generalizing the definition of divergence for the case of cogradient and contragradient vectors for a manifold of n dimensions, we can write the equation:

$$\operatorname{div} \mathbf{a}^i = \frac{\partial a_l}{\partial x_l}, \quad \operatorname{div} \mathbf{a}_i = \frac{\partial a_l}{\partial x_l} \tag{92}$$

Such sums will be affine scalars; in fact, for affine orthogonal transformations we have:

$$\overline{a}_l = \alpha^l_\beta a_\beta$$

Also we have:

$$\frac{\partial}{\partial \overline{x}_l} = \frac{\partial}{\partial x_\sigma} \cdot \alpha^l_\sigma.$$

Therefore:

$$\frac{\partial \overline{a}_l}{\partial \overline{x}_l} = \alpha^l_\sigma \frac{\partial x^l_\beta \, a_\beta}{\partial x_\sigma} = \alpha^l_\sigma \, \alpha^l_\beta \frac{\partial a_\beta}{\partial x_\sigma} = \delta^\beta_\sigma \frac{\partial a_\beta}{\partial x_\sigma} = \frac{\partial a_\sigma}{\partial x_\sigma},$$

which proves the invariance of div $\mathbf{a}_i$ for affine orthogonal transformations. It is obvious that in the general case the equations (92) will not be scalars and therefore the generalization of the notion of divergence should be sought in another direction.

It is obvious that instead of the ordinary derivative we should use the tensor derivative. And because the index of the variable with respect to which we take the derivative, coincides with one of the indices of the tensor and over it the sum from 1 to n is taken, then it is clear that to the tensor derivative it is necessary to apply the method of eliminating one of the indices i.e., it is necessary to apply the unit composition.

We call cogradient divergence of the tensor $\boldsymbol{T}^{i_1,i_2\ldots i_r}_{j_1,j_2\ldots j_3}$ *with respect to the cogredient index i_k the equation denoted by* $\boldsymbol{Div}\ \boldsymbol{T}^{i_1 i_2\ldots i_r}_{j_1 j_2\ldots j_s}$ *and defined by the equation:*

$$\mathbf{Div}\, T^{i_1 i_2\ldots i_r}_{j_1 j_2\ldots j_s} = \delta^l_\sigma\, T^{i_1 i_2\ldots i_{k-1}\sigma i_{k+1}\ldots i_r}_{j_1 j_2\ldots j_s \cdot l} = T^{i_1 i_2\ldots i_{k-1} l i_{k+1}\ldots i_r}_{j_1 j_2\ldots j_s \cdot l} \tag{93}$$

The cogradient divergence of any tensor will be obviously a tensor with a rank that is lower by 1 than the rank of the initial tensor; this lowering of the rank affects the cogradient indices; so the cogradient divergence can be formed only for a tensor which has at least one cogradient index. Obviously we cannot form cogradient divergence of a contragredient tensor. To simplify the notation we will not indicate the index with respect to which the cogradient divergence is taken; in most cases this will not lead to any misunderstanding.

In the case when the cogradient divergence produces a tensor of zeroth rank, i.e., a scalar, it will be denoted by a normal font Div not by bold font.

We will form a cogradient divergence of a cogradient vector, and also a cogradient and a mixed tensor of the second rank; we will always take Riemann tensor derivatives and will have:

$$Div\, \mathbf{a}^i = \mathbf{a}^l_{\cdot l}$$

but

$$a^i_l = \frac{\partial a^i}{\partial x_l} + \left\{ \begin{matrix} \sigma\, l \\ i \end{matrix} \right\} a^\sigma,$$

from where:

$$Div\, \mathbf{a}^i = \frac{\partial a^l}{\partial x_l} + \left\{ \begin{matrix} \sigma\, l \\ l \end{matrix} \right\} a^\sigma$$

but by formula (51) we have:

$$\left\{ \begin{matrix} \sigma\, l \\ l \end{matrix} \right\} = \frac{\partial lg\, \sqrt{g}}{\partial x_\sigma},$$

from where:

$$Div\,\mathbf{a}^i = \frac{1}{\sqrt{g}}\,\frac{\partial\sqrt{g}\,a^l}{\partial x_i}. \qquad (94)$$

By doing the similar calculations we find:

$$\mathbf{Div}\ \mathbf{T}^{ik} = Q^i = \frac{1}{\sqrt{g}}\,\frac{\partial\sqrt{g}\,T^{li}}{\partial x_l} + \left\{{\sigma\,l \atop i}\right\} T^{\sigma l}, \qquad (95)$$

if the divergence is taken with respect to the second index; if the divergence is taken with respect to the fist index then we will get the following formula:

$$\mathbf{Div}\ \mathbf{T}^{ik} = P^k = \frac{1}{\sqrt{g}}\,\frac{\partial\sqrt{g}\,T^{lk}}{\partial x_l} + \left\{{\sigma\,l \atop k}\right\} T^{l\sigma}$$

It is easy to see that for an antisymmetric tensor the second sum in formula (95) will contain terms which are eliminated in pairs. So for the antisymmetric tensor we will have:

$$\mathbf{Div}\ \mathbf{T}^{ik} = \frac{1}{\sqrt{g}}\,\frac{\partial\sqrt{g}\,T^{il}}{\partial x_l}. \qquad (96)$$

The cogradient divergence of a cogradient tensor of the second rank will be a cogradient vector; if a tensor of the second rank is an antisymmetric tensor, it is not difficult to see, that the cogradient divergence of the obtained cogradient vector will be zero:

$$Div\left(\mathbf{Div}\ \mathbf{T}^{ik}\right) = 0, \qquad (97)$$

for the antisymmetric tensor $\mathbf{T}^{ik}$. In fact, according to formula (96), we have:

$$Q^i = \frac{1}{\sqrt{g}}\,\frac{\partial\sqrt{g}\,T^{il}}{\partial x_l},$$

and:

$$Div\,Q^i = \frac{1}{\sqrt{g}}\,\frac{\partial\sqrt{g}\,Q^l}{\partial x_l} = \frac{1}{\sqrt{g}}\,\frac{\partial^2 T^{lm}}{\partial x_l \partial x_m},$$

but, as $T^{lm} = -T^{ml}$, then the sum on the right-hand side of the above equation will have terms that are eliminated in pairs and therefore will be equal to zero, which proves formula (97).

Using formulas (53) and (95) we will prove that the cogradient divergence of the conjugate of the fundamental tensor $\mathbf{g}^{ik}$ is equal to zero;

$$\mathbf{Div}\ \mathbf{g}^{ik} = 0. \qquad (98)$$

Let us form a cogradient divergence of a mixed tensor of the second rank $\mathbf{T}_l^k$; it is obvious that the divergence can be formed only with respect to the upper cogradient index and in this case this gives us a contragradient vector:

$$\mathbf{Div}\ \mathbf{T}_l^k = \mathbf{Q}_i = \frac{1}{\sqrt{g}}\frac{\partial\sqrt{g}\,T_i^l}{\partial x_l} - \begin{Bmatrix} i\,l \\ \sigma \end{Bmatrix} T_\sigma^l, \tag{99}$$

which formula is obtained with the help of transformations completely analogous to those which were used earlier.

We form from the tensor $\mathbf{T}_i^k$ a cogradient tensor T^{ik} with the help of the principal composition:

$$T^{ik} = g^{\alpha k}\,T_\alpha^i$$

and assume that the tensor, formed in such a way, will be a symmetric tensor. In such a case, formula (99) can be transformed in one of the following formulas:

$$\begin{aligned}
\mathbf{Div}\,T_i^k &= \frac{1}{\sqrt{g}}\frac{\partial\sqrt{g}\,T_i^l}{\partial x_l} - \frac{1}{2}T^{lm}\frac{\partial g_{lm}}{\partial x_l}, \\
\mathbf{Div}\,T_i^k &= \frac{1}{\sqrt{g}}\frac{\partial\sqrt{g}\,T_i^l}{\partial x_l} + \frac{1}{2}T_{lm}\frac{\partial g^{lm}}{\partial x_i},
\end{aligned} \tag{100}$$

where T_k^i is obtained from the mixed tensor also with the help of the principal composition:

$$T_{ik} = g_{\alpha k}\,T_i^\alpha.$$

We will derive formula (100). As $\begin{Bmatrix} i\,l \\ \sigma \end{Bmatrix} = g^{\sigma\alpha}\begin{bmatrix} i\,l \\ \alpha \end{bmatrix}$, then by taking into account equation (41) for for square brackets of Christoffel we find the following equation:

$$\begin{Bmatrix} i\,l \\ \sigma \end{Bmatrix} T_\sigma^l = g^{\sigma\alpha}T_\sigma^l\begin{bmatrix} i\,l \\ \alpha \end{bmatrix} = T^{l\alpha}\begin{bmatrix} i\,l \\ \alpha \end{bmatrix} = \frac{1}{2}T^{l\alpha}\left(\frac{\partial g_{i\alpha}}{\partial x_l} + \frac{\partial g_{l\alpha}}{\partial x_i} - \frac{\partial g_{li}}{\partial x_\alpha}\right),$$

but:

$$T^{l\alpha}\frac{\partial g_{i\alpha}}{\partial x_l} = T^{\alpha l}\frac{\partial g_{i\alpha}}{\partial x_l} = T^{l\alpha}\frac{\partial g_{il}}{\partial x_\alpha}.$$

Using the symmetry of the tensor $\mathbf{T}^{lk}$ we get:

$$\begin{Bmatrix} i\,l \\ \sigma \end{Bmatrix} T_\sigma^l = \frac{1}{2}T^{l\alpha}\frac{\partial g_{i\alpha}}{\partial x_i} = \frac{1}{2}T^{lm}\frac{\partial g_{lm}}{\partial x_i},$$

which proves the first of the formulas (100).

The second of the formulas (100) is derived in the following way; since:

$$T^{lm} = g^{l\alpha} T_\alpha^m$$

then:

$$T^{lm} \frac{\partial g_{lm}}{\partial x_i} = T_\alpha^m g^{l\alpha} \frac{\partial g_{lm}}{\partial x_i}$$

but $g^{l\alpha} g_{lm} = \delta_m^\alpha$, so differentiating the obtained equation with respect to x_i we find:

$$g^{l\alpha} \frac{\partial g_{lm}}{\partial x_i} + g_{lm} \frac{\partial g^{l\alpha}}{\partial x_i} = 0,$$

from where:

$$T^{lm} \frac{\partial g^{lm}}{\partial x_i} = -T_\alpha^m g_{lm} \frac{\partial g_{l\alpha}}{\partial x_i} = -g_{ml} T_\alpha^m \frac{\partial g^{\alpha l}}{\partial x_i} = -T_{\alpha l} \frac{\partial g^{\alpha l}}{\partial x_i} = -T_{lm} \frac{\partial g^{lm}}{\partial x_i},$$

which proves the second of the formulas (100).

We saw above that we cannot form cogradint divergency for a contragredient tensor. The generalization of the notion of divergence for the case of contragradient tensor leads us to the idea of contragradient divergency. For the formation of contragradient divergency, with the help of the principal composition, we raise one of the contragradient indices and on the obtained tensor apply the operation of cogradient divergency. This operation can be always performed, because the obtained tensor will have at least one cogradient index. The contragradient divergency is always with respect to one of contragredient indices of the tensor namely to this, which we raise. We will denote the contragradient divergency by the symbol **Δiv**, and in the case, if it produces a scalar, then instead of a bold font we will be write it with a normal font: Δiv

Contragradient divergency of the tensor $\boldsymbol{T}_{j_1 j_2, \ldots j_s}^{i_1 i_2 \ldots i_r}$, *with respect to the index* j_k *is called a tensor defined by the equations:*

$$\boldsymbol{\Delta iv} \ \boldsymbol{T}_{j_1 j_2 \ldots j_s}^{i_1 i_2 \ldots i_r} = \boldsymbol{Div} \ \boldsymbol{T}_{j_1 j_2 \ldots j_{k-1} j_{k+i} \ldots j_s}^{i_1 i_2 \ldots i_r \sigma}, \tag{101}$$

where the cogradient divergency is taken, with respect to the index σ, *and the tensor under the symbol* $\boldsymbol{Div}$ *is defined by the equation:*

$$T_{j_1 j_2 \ldots j_{k-1} j_{k+1} \ldots j_s}^{i_1 i_2 \ldots i_r \sigma} = g^{\alpha \sigma} T_{j_1 j_2 \ldots j_{k-1} \alpha j_{k+1} \ldots j_r}^{i_1 i_2 \ldots i_r}.$$

Let us define the contragradient divergency of the contragradient vector $\mathbf{a}_i$. We have:

$$\Delta iv \, \mathbf{a}_i = Div \, \mathbf{a}^\sigma,$$

152

where:

$$\mathbf{a}^{\sigma} = g^{i\sigma}\, a_i,$$

but:

$$Div\, a^{\sigma} = \frac{1}{\sqrt{g}}\, \frac{\partial \sqrt{g}\, a^l}{\partial x_l},$$

and for this reason:

$$\Delta iv\, \mathbf{a}_i = \frac{1}{\sqrt{g}}\, \frac{(\sqrt{g}\, g^{lm}\, a_m)}{\partial x_l}. \tag{102}$$

It is easy, by simple calculation, to find the equation:

$$\Delta iv\, \mathbf{g}_{ik} = 0. \tag{98'}$$

It is not difficult with the help of formula (102) to obtain the equation which is a generalization of the Laplace operation:

$$\frac{\partial^2 f}{\partial x_1^2} + \frac{\partial^2 f}{\partial x_2^2} + \frac{\partial f}{\partial x_2^2}.$$

In fact we take for a_i the gradient f, i.e., set $\mathbf{a}_i = grad\, f$, $a_i = \frac{\partial f}{\partial x_i}$; formula (102) gives us in this case:

$$\Delta iv\, (grad\, f) = \frac{1}{\sqrt{g}}\, \frac{\partial\left(\sqrt{g}\, g^{lm}\, \frac{\partial f}{\partial x_m}\right)}{\partial x_l}.$$

The operation under the function f which is on the right-hand side of the obtained equation is called *generalized of the operation of Laplace* or *differential parameter of the second rank*. We will denote this operation by the symbol $\square f$, in this case the previous equation will be written in the following way:

$$\square f = \Delta iv\left(grad\, f\right) = \frac{1}{\sqrt{g}}\, \frac{\partial}{\partial x_l}\left(\sqrt{g}\, g^{lm}\, \frac{\partial f}{\partial x_m}\right). \tag{103}$$

According to the formation of the contragradient divergency, the operation of Laplace will be a scalar, i.e., its form will not change when the variables of the manifold M_n change. We will remark here that the tensor calculus allows us to arrive very easily at operations, depending on the first derivatives of one or two functions and not changing its kind when the variables are changed.

It is easy to see that the expressions:

$$\Delta\varphi = g^{\alpha\beta}\,\frac{\partial\varphi}{\partial x_\alpha}\,\frac{\partial\varphi}{\partial x_\beta} \tag{104}$$

$$\Delta(\varphi\psi) = g^{\alpha\beta}\,\frac{\partial\varphi}{\partial x_\alpha}\,\frac{\partial\varphi}{\partial x_\beta}, \tag{105}$$

called *fist and mixed differential parameters* which are scalars and therefore they do not change their form under change of variables
.

If we choose g_{ik} to be constant and assume that $g_{ik} = 0$, $i \neq k$, $g_{ik} = 1$, then the operation of Laplace will have the following form:

$$\Box f = \frac{\partial^2 f}{\partial x_1^2} + \frac{\partial^2 f}{\partial x_2^2} + \ldots + \frac{\partial^2 f}{\partial x_3^2},$$

because in our case $g = 1$, $g^{ik} = \delta_i^k$. Thus the introduced operation of Laplace is a natural generalization of the classical operation of Laplace. The operation of Laplace, the first and the mixed differential parameters possess such an important property, that any scalar, expressed through the derivative of the function f with respect to $x_1, x_2 \ldots x_n$ can be obtained with the help of combination of these three operations. We do not discuss here the proof of this interesting idea, but we direct our readers to the classical book *Darboux, Leçons sur la théorie générale des surfaces*.

Using the generalization of the notion of divergency, it is not difficult to solve the problem of the transformation of the divergency of a vector in M_3 into curvilinear coordinates, where the components of the fundamental tensor are equal to 1 with equal indices and zero when the indices are different (see section 5 of the previous chapter). In this case for the divergency of vectors $\mathbf{a}^i$ or $\mathbf{b}^i$ we, according to formulas (94) and (102), will have:

$$\begin{aligned}
Div\,\mathbf{a}^i &= \frac{\partial a^1}{\partial x_1} + \frac{\partial a^2}{\partial x_2} + \frac{\partial a^3}{\partial x_3}, \\
\Delta iv\,\mathbf{a}_i &= \frac{\partial a_1}{\partial x_1} + \frac{\partial a_2}{\partial x_2} + \frac{\partial a_3}{\partial x_3}
\end{aligned} \tag{106}$$

We will denote the expressions on the right-hand side of these equations, by the usual notations: $div\,\mathbf{a}^i$ or $div\,\mathbf{a}_i$. As the cogradient and contragradient divergences are invariants, then, in any $\overline{M}_3$, obtained from M_3 with the help of point transformations, we will have:

$$div\,\mathbf{a}^i = Div\,\mathbf{a}^i = \frac{1}{\sqrt{\overline{g}}}\,\frac{\partial\sqrt{\overline{g}}\,\overline{a}^l}{\partial\overline{x}_l},$$

154

$$div\, \mathbf{a}_i = \Delta iv\, \mathbf{a}_i = \frac{1}{\sqrt{\overline{g}}}\, \frac{\partial \sqrt{\overline{g}}\, \overline{g}^{lm}\, \overline{a}_m}{\partial \overline{x}_l}.$$

We turn to the case of orthogonal curvilinear coordinates, i.e., replacing $\overline{M}_3$ by $M_3^{(q)}$, and replacing $\overline{g}_{ik}, \overline{g}^{ik}$ by the following expressions obtained in the previous chapter:

$$\overline{g}_{ii} = H_i^2, \quad \overline{g}_{ik} = 0, \quad i \neq k$$

$$\sqrt{\overline{g}} = H_1\, H_2\, H_3\, \overline{g}^{ik} = \frac{1}{H_i^2}, \quad g^{ik} = O, i \neq k$$

we find the following formulas for the transformation of divergence in curvilinear coordinates;

$$div\, \mathbf{a}^1 = \frac{1}{H_1 H_2 H_3}\left\{ \frac{\partial H_1 H_2 H_3 \overline{a}^1}{\partial q_1} + \frac{\partial H_1 H_2 H_3 \overline{a}^2}{\partial q_2} + \frac{\partial H_1 H_2 H_3 \overline{a}^3}{\partial q_3} \right\}$$

$$div\, \mathbf{a}^1 = \frac{1}{H_1 H_2 H_3}\left\{ \frac{\partial \frac{H_2 H_3}{H_1}\overline{a}^1}{\partial q_1} + \frac{\partial \frac{H_3 H_1}{H_2}\overline{a}^2}{\partial q_2} + \frac{\partial \frac{H_1 H_2}{H_3}\overline{a}^3}{\partial q_3} \right\}$$

Expressing a^i through $a^{(q_i)}$ and a_i through $a_{(q_i)}$ and using (105), we find:

$$div\, \mathbf{a}^i = \frac{1}{H_1 H_2 H_3}\left\{ \frac{\partial H_2 H_3 a^{(q_1)}}{\partial q_1} + \frac{\partial H_3 H_1 a^{(q_2)}}{\partial q_2} + \frac{\partial H_1 H_2 a^{(q_3)}}{\partial q_3} \right\}$$

$$div\, \mathbf{a}_i = \frac{1}{H_1 H_2 H_3}\left\{ \frac{\partial H_2 H_3 a_{q_1}}{\partial q_1} + \frac{\partial H_3 H_1 a_{q_2}}{\partial q_2} + \frac{\partial H_1 H_2 a_{q_3}}{\partial q_3} \right\}$$

These equations show that the divergence both cogradient and contragradient is expressed in the same way, with the help of the projection of the vector on the directions q_i of systems of orthogonal curvilinear coordinates.

As the Laplace operator $\Box f$ is a divergence of the contragredient vector of gradient, then for the transformation of the Laplace operator to curvilinear coordinates we will have the following formula:

$$\Box f = \frac{1}{H_1 H_2 H_3}\left\{ \frac{\partial \frac{H_2 H_3}{H_1}\frac{\partial f}{q_1}}{\partial q_1} + \frac{\partial \frac{H_3 H_1}{H_2}\frac{\partial f}{q_2}}{\partial q_2} + \frac{\partial \frac{H_1 H_2}{H_3}\frac{\partial f}{q_3}}{\partial q_3} \right\} \qquad (107)$$

2. As an application of the notions of cogradient and contragredient divergence, we will consider the proof that the contragredient divergence of some tensor is equal to zero, being similar to the contraction of the Riemannian symbol R_{ik}

We will prove that the divergence of the tensor:

$$B_{ik} = R_{ik} - \frac{1}{2} g_{ik} R$$

is equal to zero; for $n = 4$ the tensor B_{ik} coincides with the tensor T_{ik} introduced with the help of formula (66).

We will prove the equation:

$$\Delta i v \mathbf{B}_{ik} = \Delta i v \left(\mathbf{R}_{ik} - \frac{1}{2} \mathbf{g}_{ik} R \right) = 0. \tag{108}$$

For the simplification of the calculations it should be noticed that $\Delta i v$, in the left-hand side of equation (108) is a contragradient vector, so if we prove that all of its components are equal to zero for any specially chosen variables, then by the property of a vectors *for any variables* the components of this contragradient vector will be zero

Considering equation (108) with the given values of the variables x_i we can choose a system of variables which is the simplest for the given values of the variable. Of course we should be very careful with the process of differentiating and introduce the simplification only after the differentiation is carried out.

Tensor derivatives, curvature, Riemannian symbols and so on, can be significantly simplified if we replace the variables x_i with such new variables $\overline{x}_i$ that for the given values $x_i = x_i^{(0)}$, all tensorial parameters will be zero. And of course it does not mean at all that the derivatives of these parameters will be zero, because the parameters become zero only for $x_i = x_i^{(0)}$. We will show how we can replace the variables in the case of symmetric tensorial parameters in order that $\overline{\Gamma}^I_{\lambda\mu}$ will be zero for $x_i = x_i^{(0)}$

We will introduce new variables x_i under the following conditions:

$$x_i = x_i^{(0)} + \overline{x}_i - \frac{1}{2} \left(\Gamma^i_{\alpha\beta} \right) \overline{x}_\alpha \, \overline{x}_\beta,$$

wherein $(\Gamma^i_{\alpha\beta})_0$ is taken for the value $x_i = x_i^{(0)}$, i.e., it is constant. It is obvious that the values $x_i = x_i^{(0)}$ correspond to the values $\overline{x}_i = 0$. We will show that for $\overline{x}_i = 0$ and so for $x_i = x_i^{(0)}$ the values of transformation of tensorial parameters $\overline{\Gamma}^i_{\lambda\mu}$ will be zero.

We have $\overline{x}_i = O$:

$$\left(\frac{\partial x_\nu}{\partial \overline{x}_i} \right)_0 = \delta^\nu_i, \quad \left(\frac{\partial^2 x_\nu}{\partial \overline{x}_\lambda \partial \overline{x}_\mu} \right) = -\frac{1}{2} \left(\Gamma^\nu_{\lambda\mu} \right)_0 - \frac{1}{2} \left(\Gamma^\nu_{\mu\lambda} \right)_0 = -\left(\Gamma^\nu_{\lambda\mu} \right),$$

156

by the symmetry of tensorial parameters $\Gamma^i_{\lambda\nu}$; inserting) the obtained expressions for the derivatives in formula (28)

$$\frac{\partial x_\nu}{\partial x_i} \quad \text{and} \quad \frac{\partial^2 x_\nu}{\partial \overline{x}_\lambda \partial \overline{x}_\mu}$$

when $\overline{x}_1 = 0$ we find the equation:

$$-\left(\Gamma^\nu_{\lambda\mu}\right)_0 = \left(\overline{\Gamma}^i_{\lambda\mu}\right)_0 \delta^\nu_i - \left(\Gamma^\nu_{\alpha\beta}\right)_0 \delta^\alpha_\lambda \delta^\beta_\mu,$$

from where;

$$\left(\overline{\Gamma}^\nu_{\lambda\mu}\right)_0 = 0$$

when $\overline{x}_i = 0$, i.e., when $x_i = x_i^{(0)}$ and for all ν, λ, μ. In such a way by using the already introduced transformation of variables, we set to zero all tensorial parameters.

Similar kind of transformations can be carried out because the tensorial parameters are not tensors; in one system of variables all of their components can be zeros, in another – no. The property of symmetry of tensorial parameters is necessary in order that by transformations of variables we can make them zero. In fact, replacing x_i by $\overline{x}_i$ in formula (27) we find:

$$\Gamma^i_{\lambda\mu} = \overline{\Gamma}^\sigma_{\alpha\beta} \frac{\partial \overline{x}_\alpha}{\partial x_\lambda} \frac{\partial \overline{x}_\beta}{\partial x_\mu} \frac{\partial x_i}{\partial x_\sigma} + \frac{\partial^2 \overline{x}_\sigma}{\partial x_\lambda \partial x_\mu} \frac{\partial x_i}{\partial \overline{x}_\sigma},$$

if we can, with the help of transformation of the variables, convert all tensorial parameters into zeros $\left(\overline{\Gamma}^\sigma_{\alpha\beta} = 0\right)$, then for the initial tensorial parameters we will have the equation:

$$\Gamma^i_{\lambda\mu} = \frac{\partial x_i}{\partial \overline{x}_\sigma} \frac{\partial^2 \overline{x}_\sigma}{\partial x_\lambda \partial x_\mu}.$$

This equation show, that $\Gamma^i_{\lambda\mu}$ should be symmetric parameters.

Because the curly brackets of Christoffel are symmetric parameters, then we can always find such transformation of variables that for $x_i = x_i^{(0)}$ we have:

$$\left\{ \begin{matrix} i\,k \\ l \end{matrix} \right\} = 0$$

We call such a system of variables *zeroth system* for the values $x_i = x_i^{(0)}$

For the zeroth system of variables by formulas (48), (49), (51), (52), and (56) such equations will follow (which hold only when $x_i = x_i^{(0)}$).

$$\begin{bmatrix} i\,k \\ l \end{bmatrix} = 0,$$

$$\frac{\partial g_{ik}}{\partial x_l} = 0, \quad \frac{\partial g^{ik}}{\partial x_l} = 0, \quad \frac{\partial lg\,\sqrt{g}}{\partial x_l} = 0, \qquad (109)$$

$$(ki, lm) = \frac{1}{2}\left(\frac{\partial^2 g_{km}}{\partial x_i \partial x_l} + \frac{\partial^2 g_{il}}{\partial x_k \partial x_m} - \frac{\partial^2 g_{kl}}{\partial x_i \partial x_m} - \frac{\partial^2 g_{im}}{\partial x_k \partial x_l} \right)$$

We will prove that besides the last of the formulas for (ki, lm) there exists one more relation (when $x_i = x_i^{(0)}$):

$$\frac{\partial\,(ki, lm)}{\partial x_s} = \frac{1}{2}\left(\frac{\partial^3 g_{km}}{\partial x_s \partial x_i \partial x_l} + \frac{\partial^3 g_{il}}{\partial x_s \partial x_k \partial x_m} - \frac{\partial^3 g_{kl}}{\partial x_s \partial x_i \partial x_m} - \frac{\partial^3 g_{im}}{\partial x_s \partial x_k \partial x_l} \right).$$
$$(110)$$

It is clear that the above equation should not be obtained, as we discussed above, by ordinary differentiation of the last of formulas (109), but it is necessary to use the relation (56). Differentiating it with respect to x_s, we notice that the first four terms will give us the expression in formula (110) when differentiating the remaining terms one of the square brackets of Christoffel will always be a multiplier, which becomes zero when $x_i = x_i^{(0)}$; in such a way formula (110) will be proved.

We will prove now formula (108). With the help of the principal composition we will form the tensor B_k^l.

$$B_k^l = g^{\alpha l} B_{\alpha k} = g^{\alpha l} R_{\alpha k} - \frac{1}{2} g^{\alpha l} g_{\alpha k} R = g^{\alpha l} R_{\alpha k} - \frac{1}{2} \delta_k^l R$$

and by the definition of the contragradient divergence we have:

$$\Delta iv\, \mathbf{B}_{ik} = \mathbf{Div}\, \mathbf{B}_k^\sigma = \frac{1}{\sqrt{g}} \frac{\partial \sqrt{g}\, B_k^l}{\partial x_l} - \begin{Bmatrix} k\,l \\ \sigma \end{Bmatrix} B_\sigma^{l\cdot} =$$

$$= \frac{\partial B_k^l}{\partial x_l} + B_k^l \frac{\partial lg\,\sqrt{g}}{\partial x_l} - \begin{Bmatrix} k\,l \\ \sigma \end{Bmatrix} B_\sigma^l.$$

Inserting in the first term of the first part of this equation the obtained above expression for B_k^l we find:

$$\Delta iv\, \mathbf{B}_{ik} = \frac{\partial g^{\alpha l} R_{\alpha k}}{\partial x_l} - \frac{1}{2} \frac{\partial \delta_k^l R}{\partial x_l} + B_k^l \frac{\partial lg\,\sqrt{g}}{\partial x_l} - \begin{Bmatrix} k\,l \\ \sigma \end{Bmatrix} B_\sigma^l =$$

$$= \frac{\partial g^{\alpha l} R_{\alpha k}}{\partial x_l} - \frac{1}{2} \frac{\partial R}{\partial x_k} + B_k^l \frac{\partial lg\,\sqrt{g}}{\partial x_l} - \begin{Bmatrix} k\,l \\ \sigma \end{Bmatrix} B_\delta^l,$$
$$(*)$$

158

but by the definition of $R_{\alpha k}$ and by formula (68) for R, we will have:

$$R_{\alpha k} = g^{\tau\sigma}\,(\alpha\sigma, \tau k),$$

$$R = g^{\alpha l}\,g^{\tau\sigma}\,(\alpha\sigma, \tau l).$$

Inserting these values in the formula for $\mathbf{\Delta iv}\,\mathbf{B}_{ik}$ we find:

$$\Delta iv\,\mathbf{B}_{ik} = g^{\alpha l}\,g^{\tau\sigma}\left\{\frac{\partial(\alpha\sigma, \tau k)}{\partial x_l} - \frac{1}{2}\frac{\partial(\alpha\sigma, \tau l)}{\partial x_k}\right\} + (\alpha\sigma, \tau k)\frac{\partial(g^{\alpha l}g^{\tau\sigma})}{\partial x_l} -$$

$$= \frac{1}{2}(\alpha\sigma, \tau l)\frac{\partial(g^{\alpha l}\,g^{\tau\sigma})}{\partial x_k} + B_k^l\frac{\partial lg\sqrt{g}}{\partial x_l} - \left\{{k\;l}\atop{\sigma}\right\}B_\sigma^l,$$

Going to the zeroth system of variables and using formulas (109) and (110) we find that for $x_i = x_i^{(0)}$ the following relation will hold:

$$\Delta iv\mathbf{B}_{ik} = \frac{1}{2}g^{\alpha l}\,g^{\tau\sigma}\left\{\frac{\partial^3 g_{\alpha k}}{\partial x_l\partial x_\sigma\partial x_\tau} + \frac{\partial^3 g_{\sigma\tau}}{\partial x_l\partial x_\alpha\partial x_k} - \frac{\partial^3 g_{\alpha\tau}}{\partial x_l\partial x_\tau\partial x_k} - \frac{\partial^3 g_{\sigma k}}{\partial x_l\partial x_\alpha\partial x_\tau} - \right.$$

$$\left. - \frac{1}{2}\left(\frac{\partial^3 g_{\alpha l}}{\partial x_k\partial x_\sigma\partial x_\tau} + \frac{\partial^3 g_{\sigma\tau}}{\partial x_k\partial x_\alpha\partial x_l} - \frac{\partial^3 g_{\alpha\tau}}{\partial x_k\partial x_\sigma\partial x_l} - \frac{\partial^3 g_{\sigma l}}{\partial x_k\partial x_\alpha\partial x_\tau}\right)\right\}$$

It is easy to see that we can write the following equations:

$$g^{\alpha l}\,g^{\tau\sigma}\frac{\partial^3 g_{\alpha k}}{\partial x_l\partial x_\sigma\partial x_\tau} = g^{\alpha l}\,g^{\tau\sigma}\frac{\partial^3 g_{\sigma k}}{\partial x_l\partial x_\alpha\partial x_\tau}, \qquad (**)$$

$$g^{\alpha l}\,g^{\tau\sigma}\frac{\partial^3 g_{\sigma\tau}}{\partial x_l\partial x_\alpha\partial x_k} = \frac{1}{2}g^{\alpha l}\,g^{\tau\sigma}\frac{\partial^3 g_{\alpha l}}{\partial x_k\partial x_\sigma\partial x_\tau} + \frac{1}{2}g^{\alpha l}\,g^{\tau\sigma}\frac{\partial^3 g_{\sigma\tau}}{\partial x_k\partial x_\alpha\partial x_l},$$

$$g^{\alpha l}\,g^{\tau\sigma}\frac{\partial^3 g_{\alpha\tau}}{\partial x_l\partial x_\sigma\partial x_k} = \frac{1}{2}g^{\alpha l}\,g^{\tau\sigma}\frac{\partial^3 g^{\alpha\tau}}{\partial x_k\partial x_\sigma\partial x_l} + \frac{1}{2}g^{\alpha l}\,g^{\tau\sigma}\frac{\partial^3 g_{\sigma l}}{\partial x_k\partial x_\alpha\partial x_\tau}.$$

The first of these equations becomes obvious, as long as the symmetrical indices α and τ are replaced by σ and l and vice-versa in the right-hand side of the first equation. Replacing the indices α and l by σ and τ and vice-verse in the first term of the right-hand side of the second equation, we will show that it is equal to the second term, but the second term is half of the sum in the left-hand side of the second equation. In such a way we will prove the second equation and by an analogous method we can prove the third equation.

The three equations (**) show that the right-hand side of the relation, expressing $\Delta iv\,\mathbf{B}_{ik}$ in the zeroth system is equal to zero, whereas $\Delta iv\,\mathbf{B}_{ik}$ is a vector, then therefore the relation (108) holds for all systems of variables.

As the tensor $\mathbf{B}_{ik}$ obviously is a symmetric tensor, because $\mathbf{R}_{ik}$ and $\mathbf{g}_{ik}$ are symmetric tensor, it does not matter to what index we will relate $\Delta iv\,\mathbf{B}_{ik}$ to the first or the second. In both cases $\Delta iv\,\mathbf{B}_{ik}$ will be zero.

When $n=4$, $\mathbf{B}_{ik}=\mathbf{T}_{ik}$ the equation $\Delta iv\,\mathbf{T}_{ik}=0$ gives four ($n=4$) relations, four identities; when we study the law of of conservation of energy in its general form, we will see what role these identities will play.

In conclusion we will express explicitly the formula for $\Delta iv\,\mathbf{B}_{ik}$ through the tensor $\mathbf{R}_{ik}$:

$$\Delta iv\,\mathbf{B}_{ik} = g^{\alpha l}\left(\frac{\partial R_{\alpha k}}{\partial x_l} - \frac{1}{2}\frac{\partial R_{\alpha l}}{\partial x_k}\right) - g^{\sigma\tau}\left\{\begin{matrix}\sigma\,\tau\\\alpha\end{matrix}\right\}R_{\alpha k} = 0. \quad (111)$$

Using formula (*) for $\Delta iv\,\mathbf{B}_{ik}$ and noticing that $R = g^{\alpha\beta}R_{\alpha\beta}$ we obtain the equation:

$$\Delta iv\,\mathbf{B}_{ki} = g^{\alpha l}\left(\frac{\partial R_{\alpha k}}{\partial x_l} - \frac{1}{2}\frac{\partial R_{\alpha l}}{\partial x_k}\right) + R_{\alpha k}\left(\frac{\partial g_{\alpha l}}{\partial x_l} + g^{\alpha l}\frac{\partial lg\,\sqrt{g}}{\partial x_l}\right) -$$

$$- R_{\alpha\beta}\left(\frac{1}{2}\frac{\partial g^{\alpha\beta}}{\partial x_k} + \left\{\begin{matrix}k\,l\\\beta\end{matrix}\right\} - \frac{1}{2}R\left(\frac{\partial lg\,\sqrt{g}}{\partial x_k} - \left\{\begin{matrix}k\,l\\l\end{matrix}\right\}\right)\right)$$

Taking into account formulas (51), (52) and (53), we find:

$$\Delta iv\,\mathbf{B}_{ik} = g^{\alpha l}\left(\frac{\partial R_{\alpha k}}{\partial x_l} - \frac{1}{2}\frac{\partial R_{\alpha l}}{\partial x_k}\right) - g^{\sigma\tau}\left\{\begin{matrix}\sigma\,\tau\\\alpha\end{matrix}\right\}R_{\alpha k} +$$

$$+ \frac{1}{2}R_{\alpha\beta}\left(\left\{\begin{matrix}k\,\sigma\\\beta\end{matrix}\right\}g^{\sigma\beta} - \left\{\begin{matrix}k\,\sigma\\\beta\end{matrix}\right\}g^{\sigma\alpha}\right).$$

As $R_{\alpha\beta} = B_{\beta\alpha}$ then the last sum contains members which eliminate each other in pairs and therefore equal to zero; then the previous formula becomes equation (111).

3. We will define the notion of n-th fold integral in the manifold of n dimensions of M_n. We will separate from M_n the set of values of variables such that some of their functions $f(x_1, x_2, \ldots x_n)$ will have for any values of these variables a negative value or will be equal to zero. In such a way for the values of variables belonging to the chosen set there exists the following condition:

$$f(x_1, x_2, \ldots x_n) \leq 0$$

We call such kind of set of variables *area or volume V* of the manifold M_n, bounded by the *hypersurface S*, understanding by "hypersurface", only conditionally, a set of the values of variables for which

160

there exists the equation:

$$f(x_1, x_2 \ldots x_n) = 0$$

If we have a manifold of three dimensions and if we represent the variables of our manifold as rectangular coordinates, then the volume V will be ordinary volume or an area of space, bounded by the ordinary surface S, whose equation will be expressed by the equation:

$$f(x_1, x_2, x_3) = 0$$

We chose in the area V some given element of the manifold M_n: $(x_i^{(0)})$. We form the elements of the manifold, the i-coordinate in which differs by an integer (positive, equal to zero, or negative) from the quantity Δx_i, where n increments $\Delta x_1 \Delta x_2 \ldots \Delta x_n$ will be given now. We will change the numbering of all elements of the manifold M_n, which lie inside the area V. Let the number of these elements be $Q+1$ and let the j-th? of these elements has coordinates: $x_1^{(j)}, x_2^{(j)}, \ldots x_n^{(j)}$. We will form the following sum:

$$J' = \sum_{j=0}^{\Omega} F\big(x_1^{(j)}, x_2^{(j)}, \ldots x_n^{(j)}\big)\, \Delta x_1\, \Delta x_2 \ldots \Delta x_n.$$

By taking into account known analytical conditions imposed on the function F, the sum we formed above will approach the limit, if Δx_i will approach zero, whereas the number Ω will increase indefinitely. This limit, which does not depend (under known analytical conditions imposed on the function F) on the way how each of the increments Δx_i approaches zero, is called *integral from the function F on the area V (or on the volume V) and is denoted by:*

$$J = \int F(x_1, x_2, \ldots x_n)\, dV$$

where dV denotes $dx_1, dx_2 \ldots dx_n$.

The calculation of this type of integral can be done by the calculation of n simple integrals.

We see how the integral J_i changes when some variables are replaced by others.

We will replace the variables x_i by $\overline{x}_i$ – then the volume of integration V will go into some new volume V, the function $F(x_1, x_2 \ldots x_n)$ under the integral will go into a new function $\overline{F}(\overline{x}_1, \overline{x}_2 \ldots \overline{x}_n)$. According to the general theorems of integral calculus $dV = dx_1, dx_2 \ldots dx_n$

will not go into $d\overline{V} = d\overline{x}_1 d\overline{x}_2 \ldots d\overline{x}_n$ but into $d\overline{V}$ multiplied by the determinant which we already discussed, namely by the Jacobian $x_1, x_2 \ldots x_n$ by $\overline{x}_1, \overline{x}_2 \ldots \overline{x}_n$.

Setting:

$$\frac{D(x_1 x_2 \ldots x_n)}{D(\overline{x}_1 \overline{x}_2 \ldots \overline{x}_n)} = \left\| \frac{\partial x_i}{\partial \overline{x}_j} \right\|,$$

we will have:

$$dV = \frac{D(x_1 x_2 \ldots x_n)}{D(\overline{x}_1 \overline{x}_2 \ldots \overline{x}_n)} \, d\overline{V},$$

so the integral J will go into the following integral $\overline{J}$:

$$\overline{J} = \int\limits_{\overline{v}} \overline{F}(\overline{x}_1, \overline{x}_2 \ldots \overline{x}_n) \frac{D(x_1 x_2 \ldots x_n)}{D(\overline{x}_1 \overline{x}_2 \ldots \overline{x}_n)} \, d\overline{V}$$

The last formula shows that the integral J will not be invariant, as long as the function under the integral will be a scalar. The question is what kind of property the function under the integral should have in order that the integral be invariant, i.e. that its form does not change under a change of the variable.

To answer this question it is necessary to use the fundamental determinant g; because

$$\overline{g}_{ik} = g_{\alpha\beta} \frac{\partial x_\alpha}{\partial \overline{x}_i} \frac{\partial x_\beta}{\partial \overline{x}_k},$$

then, according to the general theory of multiplication of determinants, we will have:

$$\overline{g} = g \cdot \left\| \frac{\partial x_i}{\partial \overline{x}_j} \right\|^2 = g \left(\frac{D(x_1 x_2 \ldots x_n)}{D(\overline{x}_1 \overline{x}_2 \ldots \overline{x}_n)} \right)^2$$

or

$$\sqrt{\overline{g}} = \frac{D(x_1 x_2 \ldots x_n)}{D(\overline{x}_1 \overline{x}_2 \ldots \overline{x}_n)} \sqrt{g} \tag{112}$$

It follows from formula (112) that:

$$\sqrt{\overline{g}} \, d\overline{V} = \sqrt{g} \, dV.$$

In such a way, if the function F under the integral is a product of the scalar φ and $\sqrt{g}$, then the integral J will be invariant under the transformation of variables. In fact:

$$J = \int\limits_{r} \varphi \sqrt{g} \, dV$$

162

By transforming the variables we will write J in the form:

$$\overline{J} = \int\limits_{r} \overline{\varphi} \sqrt{\overline{g}}\, d\overline{V}.$$

Since $\sqrt{\overline{g}}\, d\overline{V} = \sqrt{g}\, dV$ by the just proved assertion, as φ is a scalar it is clear that the integral $\overline{J}$, resulting from the transformation of J, has the same form and kind as the integral J.

Weyl calls the product of some scalar and $\sqrt{g}$ – *scalar density*. If we have a product of the components of a tensors and $\sqrt{g}$, then a similar product (of course a set of these products) is called *tensor density*. We will denote scalar and tensor densities by bold Gothic letters, corresponding to that letter of the Latin alphabet with which we denoted the scalar or the tensor, used for forming the scalar or tensor density. Hence we will have:

$$F\sqrt{g} = \mathfrak{F}, \quad R\sqrt{g} = \mathfrak{R}, \quad R_{ik}\sqrt{g} = \mathfrak{R}_{ik}, \quad T^{i}\sqrt{g} = \mathfrak{T}^{i},$$

where F and R are scalars, and $\mathbf{R}_{ik}$ and $\mathbf{T}^{i}$ are tensors.

It is obviously that we can perform a number of operations with the tensor (and scalar) densities such as summation, multiplication by a scalar and a tensor, forming compositions with tensors and so on. But we should not multiply tensor densities by each other. We will discuss here only the formation of divergence of tensor density of a cogradient vector. We will prove that the sum:

$$\mathfrak{T} = \frac{\partial \mathfrak{T}^{l}}{\partial x_{l}}$$

will be a scalar density, as long as $\mathfrak{T}^{l}$ is a tensor density of some cogradient vector. The expression $\mathfrak{T}$ will be denoted by $div\mathfrak{T}^{i}$ and then we will have:

$$\mathfrak{T} = div\, \mathfrak{T}^{i} = \frac{\partial \mathfrak{T}^{l}}{\partial x_{l}}. \tag{113}$$

Since $\mathfrak{T}^{i}$ is a tensor density then:

$$\mathfrak{T}^{i} = \sqrt{g}\, T^{i}$$

and by forming the cogradient divergence of the vector $\mathbf{T}^{i}$ we find the scalar $\mathbf{Div}\mathbf{T}^{i}$. Multiplying this scalar by $\sqrt{g}$ we obtain the scalar density $\sqrt{g}\, Div\mathbf{T}^{i}$. Calculating this density we have:

$$Div\, \mathbf{T}^{i} = \frac{1}{\sqrt{g}}\, \frac{\partial \sqrt{g}\, T^{l}}{\partial x_{l}} = \frac{1}{\sqrt{g}}\, \frac{\partial \mathfrak{T}^{l}}{\partial x_{l}},$$

$$\sqrt{g}\, Div\, \mathbf{T}^i = \frac{\partial \mathfrak{T}^l}{\partial x_l} = \mathfrak{T},$$

which proves our assertion with respect to the quantity $\mathfrak{T}$ in formula (113).

We give the names scalar and tensor densities to the scalars and tensors, multiplied by $\sqrt{g}$ because in many integral physical laws, it is namely these "densities" that serve as functions under integrals, which represent the "density" of different physical quantities around a given point.

A very useful quantity is the notion, introduced by Weyl, of the *weight* of a tensor and of tensor or scalar density. It will be clarified later that the multiplication of the quantity g_{ik} by a constant number λ ultimately leads to a change of the scale,, i.e., to a change of the system of units. It is significant to know to what extent the introduced quantities depend on the change of the scale.

If a tensor or a density, under multiplication of g_{ik} by λ gets a multiplier λ^e, we assume that the tensor or the density have the weight e.

It is clear that one can think of tensor formations, which do not satisfy the condition of uniformity and do not separate the multiplier λ^e when all g_{ik} are multiplied by λ. But such tensors and densities, as we see later, can hardly have geometrical or physical significance.

It is easily seen that g_{ik} has wight 1, the determinant g has weight n, $\sqrt{g}$ has weight $\frac{n}{2}$, the conjugate of the fundamental tensor g^{ik} has weight -1; in fact:

$$g^{ik} = \frac{(-1)^{i+k}\, D^{ik}}{g},$$

but the weight of B^{ik} is $n-1$, and the weight of $g-n$, i.e., the weight of g^{ik} equal to -1.

Doing the corresponding calculations[1], we can see that R_{ik} has weight equal to zero, whereas the scalar curvature R has weight -1, hence the density $R\sqrt{g}$, which we will see many times later, has weight $\frac{n-2}{2}$, when $n=4$, the weight of that density is equal to 1. Weyl points out that it is always better to consider the tensor and the density, which have wight equal to zero, then, instead of $R\sqrt{g}$, we should consider the scalar density $\sqrt{g}\,R^n$, whose wight is equal to zero, when $n=4$ this scalar density becomes $R^2\sqrt{g}$. We will see later the meaning of Weyl's remark.

---

[1]We should keep in mind, that λ is not constant, but any function of the coordinates.

4. At the end we will consider the theorem of Gauss, generalized for the case of a manifold of n dimensions. In order to make the theorem more clear we will consider first a manifold of three dimensions and for this manifold we will prove the classical theorem of Gauss, using for its proof a method which will be easy to generalize for the case of manifolds of any dimensions.

The surface S, bounding the volume V in M_2, can be expressed analytically by two methods or equations:

$$f(x_1, x_2, x_3) = 0 \qquad (*)$$

or a system of equations:

$$x_1 = x_1(u_1, u_2), \quad x_2 = x_2(u_1, u_2), \quad x_3 = x_3(u_1, u_2), \qquad (**)$$

where u_1 and u_2 are arbitrary parameters. We will call the area V for which $f < 0$ *inner* and the area for which $f > 0$, *outer* area, bounded by the surface S, The normal N to the surface S, directed from inner to the outer area of the surface S, will be called *outer normal to the surface S*. The cosines of the angles of this outer normal will be defined by the formulas:

$$\cos(N, x_i) = \frac{\dfrac{\partial f}{\partial x_i}}{\sqrt{\left(\dfrac{\partial f}{\partial x_1}\right)^2 + \left(\dfrac{\partial f}{\partial x_2}\right)^2 + \left(\dfrac{\partial f}{\partial x_3}\right)^2}} \qquad i = 1, 2, 3$$

As long as the equation of the surface given in the form$(*)$ and the formulas:

$$\cos(N, x_i) = \frac{\pm \dfrac{D(x_k\, x_l)}{D(u_1\, u_2)}}{\sqrt{\left(\dfrac{D(x_2\, x_3)}{D(u_1\, u_2)}\right)^2 + \left(\dfrac{D(x_3\, x_1)}{D(u_1\, u_2)}\right)^2 + \left(\dfrac{D(x_1\, x_2)}{D(u_1\, u_2)}\right)^2}}$$

where $i = 1, 2, 3$, and the indices i, k, l, form one of the three combinations: $1, 2, 3;\ 2, 3, 1;\ 3, 1, 2$,

From the above equations and also from the following ones, obtained by differentiation of equation $(*)$ with respect to u_1 and u_2 when in it, instead of $x_1 x_2 x_3$, we use the equations $(**)$ and in such a way we transform it in the identity:

$$\frac{\partial f}{\partial x_i} \frac{\partial x_i}{\partial u_s} = 0, \quad s = 1, 2,$$

It is easy to get such ratio:

$$\frac{\frac{\partial f}{\partial x_1}}{\frac{D(x_2\,x_3)}{D(u_1\,u_2)}} = \frac{\frac{\partial f}{\partial x_2}}{\frac{D(x_3\,x_1)}{D(u_1\,u_2)}} = \frac{\frac{\partial f}{\partial x_1}}{\frac{D(x_1\,x_2)}{D(u_1\,u_2)}} = \lambda.$$

It is obvious, that the quantity λ can become zero only when we have:

$$\frac{\partial f}{\partial x_1} = \frac{\partial f}{\partial x_2} = \frac{\partial f}{\partial x_3} = 0$$

which condition corresponds to a special point on the surface S. Assuming that the surface S does not have special points we can consider λ a quantity, which is different from zero, and therefore *not changing its sign*. Changing the sign of the u_1 then we can always consider λ positive.

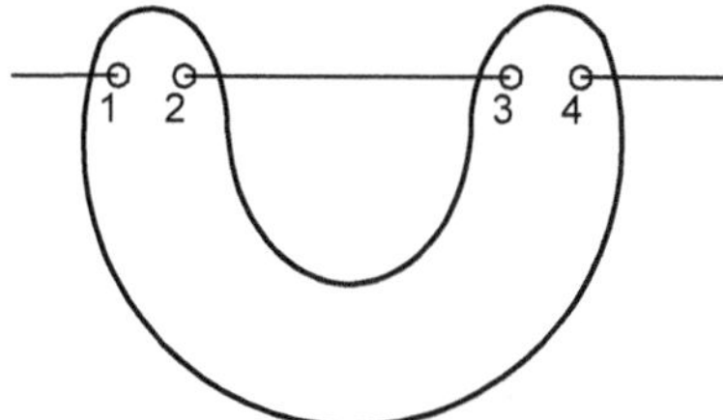

Figure 15

We will now intersect the surface S by any straight line, which is parallel to axes $x_1 : x_2 = x_2^{(0)}$, $x_3 = x_3^{(0)}$, since we assume that our surface is a closed surface, then the straight line will intersect the surface bin even number of points: (in Fig. 15 four points of intersections 1,2,3,4 are given). Moving along the straight line in such a way that x_1 is increasing then when intersecting the surface we will go from the outer to the inside area or vice-versa. The first points will be called points of *entry* (here points 1, 3), the second – points of *exit* (points 2 and 4 in Fig. 15). For the points of entry, f will decrease, $\frac{\partial f}{\partial x_1}$ will be negative; for the points of of exit, on the contrary, f will increase, so $\frac{\partial f}{\partial x_1}$ will be positive (here we assume that we stay constantly on the straight line $x_2 = x_2^{(0)}$, $x_3 = x_3^{(0)}$ which is parallel to the axis x_1). In such a way, for the points of entry we have $\frac{\partial f}{\partial x_1} < 0$, and since $\lambda > 0$ then $\frac{D(x_1 x_2)}{D(u_1 u_2)} < 0$. Also, for the points of exits we have $\frac{\partial f}{\partial x_1} > 0$ and $\frac{D(x_1 x_2)}{D(u_1 u_2)} > 0$. These analytical inequalities, which are geometrical for M_3, are quite clear, because they mean that in the points of entry $\cos(N, x_1) > 0$, i.e., the outer normal with the axis

x_1 forms an obtuse angle and in the points of exit, on the contrary, $\cos(N, x) > 0$, i.e., the outer part forms an acute angle with the axis x_1.

After these preliminary remarks we formulate the equation which is the integral theorem of Gauss for M_3:

$$\int\limits_V \frac{\partial F}{\partial x_1}\, dV = \int_S F \cos(N, x_1)\, dS$$

where dV is the volume element, and dS is the surface element S, bounding this volume. Notice, that dS has its projections on the coordinate planes of the surface elements:

$$dx_2 dx_3, \quad dx_3 dx_1, \quad dx_1 dx_2,$$

By the theorem of the theory of projection we will have:

$$(dx_2\, dx_3)^2 + (dx_3\, dx_1)^2 + (dx_1\, dx_2)^2 = dS^2,$$

or, going to the parameters u_1, u_2, we find:

$$dS = \sqrt{\left(\frac{D(x_3, x_2)}{D(u_1, u_2)}\right)^2 + \left(\frac{D(x_3, x_1)}{D(u_1, u_2)}\right)^2 + \left(\frac{D(x_1, x_2)}{D(u_1, u_2)}\right)^2}\; du_1\, du_2.$$

Remembering, that $dV = dx_1\, dx_2\, dx_3$ and doing the integration with respect to x_1 we will have the following equation:

$$\int \frac{\partial F}{\partial x_1}\, dV = -\int\limits_{S_1} F\, dx_2\, dx_3 + \int\limits_{S_2} F\, dx_2\, dx_3 - \int\limits_{S_3} F\, dx_2\, dx_2 +$$

$$+ \int\limits_{S_4} F\, dx_2\, dx_3 + \ldots + \int\limits_{S_{ik}} F\, dx_2\, dx_3$$

where S_1 is the area of the points of entry, S_2 – the area of the points of exit, adjacent to S_1, S_3 – the area of the points of entry adjacent to S_2 and so on.

But for the points of entry and exit we will have:

$$dx_2\, dx_3 = \pm \frac{D(x_2, x_3)}{(u_1, u_2)}\, du_1\, du_2,$$

where, as for the points of entry

$$\frac{D(x_2, x_3)}{D(u_1, u_2)} > 0$$

and $dx_2\,dx_3$ and $du_1\,du_2$ are positive, then it is necessary to take for points of entry the lower, and for points of exit the upper sign; therefore the previous formula can be rewritten in as:

$$\int \frac{\partial F}{\partial x_1}\,dV = \int_S F\,\frac{D(x_2, x_3)}{D(u_1, u_2)}\,du_1\,du_2 = \int_S F\,\cos(N, x_1)\,dS,$$

which proves the integral theorem of Gauss.

Everything that we said so far about the manifold M_3 can be applied word for word and to any other manifold.

The hypersurface S bounding in M_n the volume V can be expressed analytically by two methods – either by the equation:

$$f(x_1, x_2 \ldots x_n) = 0, \qquad\qquad (***)$$

or by the system:

$$x_i = x_i(u_1, u_2, \ldots u_{n-1}), \quad i = 1, 2, \ldots n, \qquad (****)$$

where $u_1, u_2, \ldots u_{n-1}$ are arbitrary parameters. The domain V will be called *inner* for variables in which $f \leq 0$, and *outer* for variables giving $f > 0$, with respect to the surface S. From equation $(***)$ and $(****)$ and analogously to the case of M_3, we introduce the relations:

$$\frac{\partial f}{\partial x_i}\,\frac{\partial x_i}{\partial x_s} = 0, \quad s = 1, 2, \ldots n - 1,$$

from where, solving these equations for $\frac{\partial f}{\partial x_i}$, we find:

$$\frac{\frac{\partial f}{\partial x_1}}{\frac{D(x_2, x_3 \ldots x_n)}{D(u_1, u_2 \ldots u_{n-1})}} = \frac{\frac{\partial f}{\partial x_2}}{\frac{D(x_3, x_4 \ldots x_n, x_1)}{D(u_1, u_2 \ldots u_{n-1})}} = \ldots = \frac{\frac{\partial f}{\partial x_n}}{\frac{D(x_1, x_2 \ldots x_n)}{D(u_1, u_2 \ldots u_{n-1})}} = \lambda$$

Here, exactly like for the case of M_3, we find that if on S there exist no values of the variables for which all $\frac{\partial f}{\partial x_i}$ are equal to zero (this case we exclude), then λ is not equal to zero, and then by a change of the sign in u_1 we can always regard λ as *positive*.

Consider now the change of f, when x_1 is changing, and

$$x_s = x_s^{(0)}(s = 2, 3, \ldots n);$$

if when x_1 is increasing, f is taking on values of variables, belonging to S, and goes from positive to negative values, then the considered system of values x_i on S is called *point of entry*, in the opposite case,

168

i.e., when from negative values f changes to positive, the system of values x_i will be called *point of exit*. For the points of entry we have $\frac{\partial f}{\partial x_1} < 0$ and by the positiveness of λ:

$$\frac{D(x_2, x_3 \ldots x_n)}{D(u_1, u_2, \ldots u_{n-1})} < 0;$$

On the contrary, for points of exit we have:

$$\frac{\partial f}{\partial x_1} > 0 \quad \text{and} \quad \frac{D(x_1, x_2, \ldots x_n)}{D(u_1, u_2, \ldots u_n)} > 0$$

Now we introduce a number of notations. We will call the expression:

$$dS = \sqrt{D} \cdot du_1 \, du_2 \ldots du_{n-1}$$

the element dS of the hypersurface S, where D is defined by the equation:

$$D = \left(\frac{D(x_2, x_3 \ldots x_n)}{D(u_1, u_2, \ldots u_{n-1})} \right)^2 + \left(\frac{D(x3, x_4, \ldots x_n,)}{D(u_1, u_2, \ldots u_{n-1})} \right)^2 + \ldots$$
$$+ \left(\frac{D(x_1, x_2, \ldots x_{n-1})}{D(u_1, u_2, \ldots u_{n-1})} \right),$$

and $\sqrt{D}$ is, of course, taken to be positive.

The above equation is obviously a direct generalization of the notion of a change of the surface S for the case M_3.

Let us define the following n cosines:

$$\cos(N, x_i) = \frac{\frac{\partial f}{\partial x_i}}{\sqrt{\left(\frac{\partial f}{\partial x_1} \right)^2 + \left(\frac{\partial f}{\partial x_2} \right)^2 + \ldots + \left(\frac{\partial f}{\partial x_n} \right)^2}}, \quad (i = 1, 2, \ldots n),$$

these equations will be denoted by one phrase, namely, saying that "*N is the direction of the outer normal*"; it is necessary to point out, that this phrase for the case $n > 3$, and especially for curved spaces does not have any geometrical meaning and expresses only the above equation defining n cosines.

Remembering that λ is a positive quantity, it is not difficult to express $\cos(N, x_i)$ by the parameters $u_1, u_2, \ldots n_{n-1}$. For example we have:

$$\cos(N < x_1) = \frac{\frac{D(x_1, x_2, \ldots x_{n-1})}{D(u_1, u_2, \ldots u_{n-1})}}{\sqrt{D}}$$

After this preliminary considerations we can prove the following integral theorem of Gauss:

Teorema 9.

$$\int\limits_V \frac{\partial F}{\partial x_i}\, dV = \int\limits_S F \cos(N, x_i)\, dS.$$

Take $i = 1$; denoting by $S_1, S_2 \ldots S_{2k-1}$ the area of points of entry on S, and by $S_2, S_4 \ldots S_{2k}$ the area of points of exit and integrating over x_1 we obtain, as in the case of M_3, the equation:

$$\int\limits_V \frac{\partial F}{\partial x_1}\, dV = -\int\limits_{S_1} F\, dx_2\, dx_3 \ldots dx_n + \int\limits_{S_2} F\, dx_2\, dx_3 \ldots dx_n + \ldots$$

$$+ \int\limits_{S_{2k}} F\, dx_2\, dx_3 \ldots dx_n.$$

In the same way as we did for M_3 we find:

$$dx_2\, dx_3 \ldots dx_n = \pm \frac{D(x_2, x_3 \ldots x_n)}{D(u_1, u_2, \ldots u_{n-1})}\, du_1\, du_2 \ldots du_{n-1}$$

where the sign $+$ is taken for the points of exit, whereas the sign $-$ for the points of entry; under this condition we can write the above integral equation as:

$$\int\limits_V \frac{\partial F}{\partial x_1}\, dV = \int\limits_S F\, \frac{D(x_2, x_3, \ldots x_n)}{D(u_1, u_2, \ldots u_{n-1})}\, du_1\, du_2 \ldots du_{n-1},$$

from where, recalling the notations for dS and for $\cos(N, x_1)$ we will prove our theorem for $i = 1$ and, therefore, we will prove it in general.

A consequence from theorem 9 follows directly and we will use it frequently later; if the function F under the integral is equal to zero on the surface S, then we have the equation:

$$\int\limits_V \frac{\partial F}{\partial x_i}\, dV = 0, \quad \text{for } F = 0 \text{ on } S. \tag{114}$$

Another consequence from the theorem of Gauss will be the formula:

$$\int\limits_V div\, \mathfrak{G}^i\, dV = \int\limits_V \frac{\partial \mathfrak{G}^l}{\partial x_l}\, dV = \int\limits_S \mathfrak{G}^{(N)}\, dS \tag{115}$$

170

where $\mathfrak{G}^{(N)}$ is defined by the formula:

$$\mathfrak{G}^{(N)} = \mathfrak{G}^i \cos(N, x_i) = \sum_{i-1}^{n} \mathfrak{G}^i \cos(N, x_i),$$

and $\mathfrak{G}^i$ is a tensor density. We will derive formula (115) from theorem 9 applying this theorem to each term in $div\,\mathfrak{G}^i$.

Formula (115) holds in this case too, if $\mathfrak{G}^i$ is not a tensor density but is a set of any n quantities $\mathfrak{G}^1, \mathfrak{G}^2, \ldots \mathfrak{G}^n$, where the symbol $div\,\mathfrak{G}^i$, which we will call *pseudo-divergence*, will be defined by the equation:

$$div\,\mathfrak{G}^i = \frac{\partial \mathfrak{G}^s}{\partial x_s} = \frac{\partial \mathfrak{G}^1}{\partial x_1} + \frac{\partial \mathfrak{G}^2}{\partial x_2} + \ldots + \frac{\partial \mathfrak{G}^n}{\partial x_n}.$$